An introduction to
Biochemistry of Fungal Development

An introduction to

Biochemistry of Fungal Development

J.E. Smith *and* D.R. Berry

Department of Applied Microbiology,
University of Strathclyde, Glasgow, Scotland

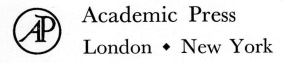

Academic Press
London • New York

A subsidiary of Harcourt Brace Jovanovich, Publishers

ACADEMIC PRESS INC., LONDON, LTD.
24–28 Oval Road
London NW1

United States Edition published by
ACADEMIC PRESS INC
111 Fifth Avenue
New York, New York 10003

Library of Congress Catalog Card Number: 73 19021
ISBN 0 12 650950 6

PRINTED IN GREAT BRITAIN BY
BUTLER AND TANNER LTD, FROME AND LONDON

Preface

All organisms proceed through a series of programmed changes in which structures are built up and transformed by complex interactions between the organism and its external environment. In this way, the organism is the product of the interaction between the information which it receives by heredity, and the resources and stimuli to which it is exposed from the environment. At a biochemical level, the relationship between gene action and macromolecular synthesis is now unfolding, and it is apparent that differentiation results from an interplay between an unchanging genome and a labile cytoplasm—the cytoplasm at any one period determining the sequence of gene expression while itself being modified as a consequence.

Many excellent treatises have appeared dealing with bacteria and higher eukaryotic organisms which indicate the complexity of the different biochemical processes which may be involved in any one developmental sequence. Regrettably, the fungi seem to have been less intensively studied in this respect than other groups and in general only a few of the many techniques available have been applied to any one system. This is particularly unfortunate since apart from the intrinsic value *per se* of studying fungi, fungi also represent an important link between the truly unicellular state and the multicellular condition of higher organisms. It can be anticipated that a fuller understanding of the factors which regulate growth and development in fungi could well be of major importance and relevance, not only to fundamental studies on development in fungi but also to the many industrial processes such as secondary metabolite production, biomass formation, biodegradation, biodeterioration, etc. in which the fungi are involved.

This present book should be useful to a wide audience range from undergraduates in bioscience to practising mycologists, plant pathologists, biochemists and chemists who want a concise reference source and introduction to this exciting and expanding aspect of biology.

The material for this book is based largely on a series of lectures given to undergraduate and postgraduate students at the University of Strathclyde. Since this book was primarily designed as an introduction to the biochemistry of fungal development we have had to omit material that would have found a place in a more extensive treatise. However, we have endeavoured to include many of the more recent review articles which will allow the student

PREFACE

to expand his knowledge relevant to his particular interests. The multidisciplinary approach to developmental studies is becoming increasingly more important and it is hoped that this book will be of value to those who do not have a background training in mycology.

We would like to thank our respective wives for their understanding and encouragement during the period of preparation, and in particular to Dr. Elizabeth Berry for her careful reading of the proofs and for the preparation of the index. We would also like to thank Miss Carol Beaumont whose artistic skills have greatly improved this book.

We are also deeply indebted to our many friends and colleagues who have assisted us by contributing original photographs of their published or unpublished works. In particular we would thank: J. G. Anderson; S. Bartnicki-Garcia; C. E. Bracker; T. D. Brock; E. C. Cantino; L. A. Casselton; E. R. Florance; G. W. Gooday; P. T. Magee; P. B. Moens; D. J. Niederpruem; E. W. Rapport and E. Streiblova.

May 1974   J. E. Smith
                                                                     D. R. Berry

Contents

CONTENTS

1 Introduction: The Conceptual Basis of Morphogenesis

1.1 Morphogenesis and Differentiation

Morphogenesis

Organisms never remain constant but proceed through a series of developmental processes of growth and reproduction which make up the life cycle. All organisms have a form; that is, they occupy space, so they can be considered to exhibit certain morphological characteristics. The form of an organism must vary throughout the life cycle even if this variation comprises only cell growth and division which occurs in the simplest organisms. In micro-organisms, total form is usually related to the form of the constituent cells and in unicellular organisms the two are synonymous. In multicellular organisms such as vertebrates, form may vary more widely, from a single fertilized egg to the complex assembly of cells, tissues and organs of the mature vertebrate. In higher organisms, the form of the whole organism is controlled more by the manner in which cells are gathered together in tissues and organs and is less dependent upon the form of individual cells. The development of an organism to the form it exhibits at any given time can be referred to as morphogenesis.

PHENOTYPE, GENOTYPE AND ENVIRONMENT

Any organism is the product of the interaction between the information which it receives by heredity, and the resources and stimuli to which it is exposed from the environment. This concept has been expressed by Darlington in the form of an equation:

$$\text{Phenotype} = \text{genotype} \times \text{environment}.$$

The genotype is the name given to that information which is contained within an organism and is passed on by heredity. The phenotype is defined as all the parameters or characteristics of an organism which can be observed, measured or studied in any way, including the chemical composition of all cellular components. Structure and morphology are one aspect of the phenotype. Whereas the genotype of a cell or organism is usually constant throughout its life cycle, the phenotype evidently varies. In view of this, it would seem apparent that the development of an organism is controlled by the environment. This is acceptable if the following conditions are stipulated:

(1) The environment is defined so as to include the "cytoplasm". "Cytoplasm" is defined in this context as all parts of the cell, excluding the genome.

(2) The composition of the cytoplasm is largely controlled by the genotype.

The relationship between the genotype, the cytoplasm and the extra-cellular environment can be illustrated by reference to simple examples. The enzyme, β-galactosidase, is produced in *Escherichia coli* when lactose is present in the medium. However, induced production of β-galacto-sidase only occurs if:

(1) The structural genes for β-galactosidase and the appropriate regulator genes are present.

(2) Lactose or a similar inducer can gain entry to the cytoplasm of the cell.

(3) There already exists in the cytoplasm a regulator protein which is sensitive to the inducer.

In micro-organisms, the metabolism of the cell can be strongly influenced by the environment, but in higher organisms the environment which exists outside the organism has negligible influence on the structure and function of the organism. As long as the environment is adequate to permit survival and growth, a species always exhibits its own characteristic form and function. The structure of a flower or the shape of a mammal are not significantly influenced by the environment. In higher organisms and particularly animals, not only does the genotype largely control the nature of the cytoplasm, it also controls the cell's external environment which consists of other cells and body fluids. Thus the very complex morphogenetic processes occurring during embryo-genesis in animals are largely insulated from the environment external to the organism.

Even in micro-organisms, it has been observed that many of the more complex morphogenetic processes become insulated from environmental change and cannot be reversed by modification of the medium after a certain stage of development, e.g., bacterial sporulation, and conidiation in *Aspergillus niger* (see Chapter 6).

In examples where morphogenetic events are clearly influenced by the environment, e.g., phytochrome responses in plants, or sexual and asexual reproduction in fungi, the environmental stimulus acts as a trigger-mechanism and the response is determined by the genotype of the organism. In plants, red light can stimulate germination of seeds, uncurling of leaves, anthocyanin production in stems, etc. Both *Aspergillus* and *Mucor* will produce asexual spores on potato extract agar, but the morphogenetic events involved are totally different in the two species.

Differentiation

Many attempts have been made to define differentiation. Most of these are valuable in that they emphasize distinctions that are useful in the study of the subject but are probably not adequate to encompass the breadth and subtlety of the concept. An Hegelian approach to the problem can therefore be justified by considering that a full appreciation of the process can only be obtained by an assimilation of a wide range of relevant information. The following statement by Gross (1968) is a valuable summation of the situation: "By definition, somatic cells of higher organisms are disposed as sets sharing structure and function. These sets are arranged as tissues, tissues as organs and organs as organisms. The sets are different from one another, hence *differentiated*." An inevitable corollary of this statement is that cellular differentiation is a necessary condition of multicellular life. If this were not so, all the cells would be of the same set, and the organism should be considered to be a colonial form of a unicellular organism as in the Parazoans.

It is doubtful, however, whether the antithesis of this is true, i.e., that differentiation cannot exist in unicellular organisms. Structures such as spores in bacteria, or asci in yeast are obviously different in form from the normal vegetative cell. Such examples, where individual cells exhibit differences in structure and function at different stages in their life cycle, have been referred to as temporal differentiation, in contrast to the more usual spatial differentiation (Bonner, 1973).

Differentiation can be studied at many different levels. Even at the descriptive level, it is not easy to establish the degree of change which can be considered to constitute differentiation. Where cells have a different shape, size or structure the distinction is usually clear. But do structural changes at a subcellular level, visible only in the electron microscope, constitute differentiation? Do changes in structure which are discernible only by the use of specific cytochemical stains constitute differentiation? It has even been proposed that the induction of a single enzyme should be considered to be differentiation!

An alternative approach has been to define differentiation with reference to the stability of the observed change. In this case, differentiation can be defined as a stable development of an altered structure and function; more transient changes being referred to as adaptation (Pasternak, 1970). On this basis, the main criterion of differentiation is the irreversibility of the process. The main disadvantage of this distinction appears to be that many processes which should be termed differentiation are reversible, at least up to a certain stage in their development, e.g., bacterial sporulation. Conversely, it is by no means certain that many processes which might well be considered to be adaptations, e.g., induction of an enzyme, are reversible. The induction process itself may be reversed, but the enzyme product usually remains.

One of the major problems in a study of differentiation is to establish the point at which the process is initiated. The appearance of a change in structure or function is merely the final stage of a continuing process, since the biosynthetic pathways and the enzymes involved must have been established previously. Even new enzyme induction is not totally satisfactory as a zero time for differentiation since the cellular changes which resulted in the induction of these enzymes would also appear to be a critical stage in the process. A zero time can more easily be established for a process of differentiation which is triggered by a change in the environment external to the organism. Unfortunately in many examples, such a clear-cut trigger is not readily discernible. During ontogeny in higher animals, the cells which give rise to different tissues and organs are delimited at a very early stage. This has been referred to as cell determination rather than differentiation. This process of determination appears to involve an irreversible change in the cell which not only persists for its lifetime but is passed on to subsequent generations of cells. It is frequently only in the latter that visible differentiation takes place. To identify the ultimate controlling mechanism for cell determination, it is necessary to look for a trigger mechanism in the early stages of blastula formation.

In this book, we have restricted our interest to those processes of development and differentiation which give rise to visible changes in fungal structure, although inevitably in the discussion of these topics relevant changes at the molecular level must be included.

"*Quis Custodiet Ipsos Custodes?*"
(Juvenal)

1.2   Biochemistry of Morphogenesis and Differentiation

Introduction

All changes in cell structure can be attributed to differences in the chemical composition of the cell either in terms of the presence or absence of cell constituents or in terms of changes in their relative abundance. It is probable that differences in the manner in which cells aggregate can be attributed to chemical differences at the cell surface. The problem of morphogenesis may be stated therefore as the means by which this change in chemical composition may be controlled.

Most organisms can exist on a relatively small number of nutrients; a carbon source, a nitrogen source, phosphorus and sulphur, metals, and in some species vitamins and amino acids. The details of the biochemical pathways which convert these nutrients into complex proteins, nucleic acids, lipids and carbohydrates and other compounds of which the cell is composed are out of place in this book. However, the mechanisms that determine which of these biosynthetic pathways are active at a given stage in cellular development are relevant to the problem of morphogenesis and differentiation. The mechanisms responsible for differentiation must be able to control the nature and quantity of the metabolites produced, the time at which they are produced and their spatial location within the cell (Bonner, 1973).

Biosynthetic reactions can be divided essentially into one of two types:

(1) The structure and quantity of the end-product of a biosynthetic reaction can be controlled by the specificity of the enzymes involved, e.g., carbohydrates, lipids, bases, amino acids and other small molecules.

(2) The structure of the end-product may be determined more or less directly by the genetic material and be independent of the enzymes involved in its biosynthesis, e.g. proteins and nucleic acids. Specialized mechanisms of quantitative regulation exist for these biosynthetic processes. Molecules synthesized in this way have been referred to as informational macromolecules (Fig. 1.1).

Biological molecules of the first type are synthesized by a series of chemical reactions, each of which is controlled by a specific enzyme. The rate of synthesis in such a biosynthetic pathway is controlled by the amount and the specific activity of the enzymes involved, the availability of substrates, and the existence in the cellular environment of molecules which stimulate or inhibit the different enzymes. It is possible that in some differentiating systems, the control of biosynthetic reactions at this level is critical (Gustafson and Wright, 1972). However, such control mechanisms are usually readily reversible and offer only a transient method of control which is not very suitable for many of the changes involved in differentiation.

The biosynthesis of most molecules is dependent upon the availability of energy. This is available in the cell as ATP which is produced by respiration, photosynthesis and other autotrophic or chemotrophic methods. Although the supply of energy may be important in controlling the presence or absence of growth and morphogenesis, it is not

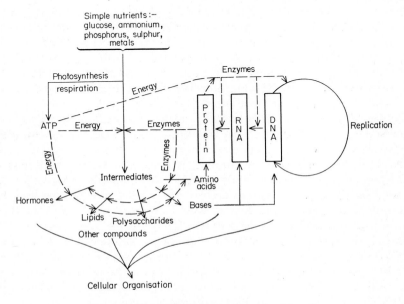

Fig. 1.1. Biosynthetic interactions in the cell.

normally the critical factor in determining which particular biosynthetic pathways are active.

In the majority of the examples of morphogenesis and differentiation studied, changes in the enzyme composition of the cell also occur. These are under direct genetic control. In examples where the first type of control is in operation, its limits are determined by the quantity and kinetic properties of the enzymes involved in the biosynthetic pathway. These are again under genetic control.

The problem of morphogenesis and differentiation can be restated in terms of the mechanism by which genetic information contained in DNA controls not only the structure of all enzymes produced in a cell but also the time at which they are produced.

Regulation of Enzyme Activity

There are basically three mechanisms for controlling enzyme activity: isosteric inhibition, allosteric control and the activation of enzyme precursors. Isosteric inhibition occurs when the inhibitor is sterically related to a substrate or a cofactor of the enzyme and competes with either at the active site. Allosteric control occurs when a low molecular weight compound acts on an enzyme at a site other than the catalytic site, causing a stimulation or inhibition in activity of the enzyme. Allosteric control occurs most frequently in enzymes which are composed of several protein sub-units and can be attributed to changes in the quaternary structure of the enzyme caused by the allosteric effector. Allosteric enzymes frequently catalyse the initial reactions of a biosynthetic pathway which is subject to feed-back inhibition. The product of such a pathway may have little or no structural similarity to the substrates of the enzyme involved in the first stages of its biosynthesis. Inhibition of these enzymes by the product must therefore be of an allosteric nature.

Enzymes may exist in the cell in an inactive state, thus requiring chemical modification before gaining their normal activity. Glycogen phosphorylase occurs in an inactive form b, which requires phosphorylation to convert it to the active form a. Trypsinogen and pancreatic lipase are activated by proteolytic enzymes. Other enzymes have been shown to be activated by glutathione, heat treatment and other stimuli. In bacteria, there is a three-minute lag between the appearance of the inducible enzyme histidinase as measured by antigenic techniques and the appearance of the enzyme activity. The enzyme is therefore first produced in an inactive form.

Isoenzymes

In many organisms, it has been demonstrated that an enzyme can exist in several different forms, called isoenzymes. Isoenzymes may differ in their amino acid composition (indicating that they are derived from different genes), in their conformation, in their subunit composition if more than one type of subunit is present, and in the total number of subunits present per enzyme molecule. They have been described for a large number of enzymes, e.g., lactate and malate dehydrogenases, aspartate aminotransferase, alkaline phosphatase and hexokinase. They appear to be widely distributed in nature and may vary between different tissues in the same organism and even between different developmental stages of the same cell. Some of the differences can be attributed to the presence of mitochondrial or plastid forms of an enzyme but the function of others is less obvious (De Vellis, 1970).

Biosynthesis of Informational Macromolecules

The majority of recent studies in cell biology have been carried out on prokaryotes, hence the control mechanisms operating in these organisms are much more fully documented than those occurring in eukaryotes. In the remainder of this chapter, the known information pertaining to the control of biosynthesis of informational macromolecules is surveyed and its relevance to morphogenesis and differentiation discussed. Most of the work referred to has been extensively reviewed elsewhere so detailed references will not be given.

DNA STRUCTURE AND FUNCTION

Prokaryotes. The genetic material has been shown to be DNA in bacteria and most bacterial viruses. The evidence indicates that this is also true in higher organisms. *In those bacteria which have been studied, the genome consists of a single circular DNA molecule.* The genetic information is contained in the sequence of bases on one of the strands. Thus, when we refer to the gene, we are in fact concerned with the sequence of nucleotides in the DNA. An analogy can be drawn between the DNA molecules and the punched tape used for programming computers; the tape itself is simply an inert carrier of the information which is contained in the sequence of holes in the tape. In most organisms the genotype remains constant during vegetative development and asexual reproduction, and the DNA composition of the cells remains either constant or quantitatively rather than qualitatively, changed. In bacteria, the concept that the genotype remains constant throughout the life cycle requires some qualification. Sexuality, colicin production and several other bacterial characteristics are controlled by episomes. These are genetic elements which can exist either in an autonomous state when they can pass from cell to cell by infection or in an integrated state when they constitute part of the bacterial chromosome. Some bacteriophage exist as episomes and exist in the integrated, lysogenic, state or the autonomous lytic state. A considerable amount of information is available on the molecular events occurring when a lysogenic phage is induced into the lytic state. Although many workers would not accept phage induction in bacteria as a process of differentiation, studies in such systems have certainly provided a valuable model system for studying developmental processes.

Quantitative changes in the level of DNA in bacteria also occur as a result of changes in the number of copies of the single chromosome in the cell. These changes do not appear to have any regulatory function and are normally brought about by changes in the growth rate.

Eukaryotes. The DNA content of eukaryotic cells is much higher than in prokaryotes varying from 10 × the prokaryote level in fungi to 1000 × in the more complex animals and plants. In eukaryotes, the DNA is divided between several different chromosomes. In addition to DNA, these chromosomes contain large amounts of basic proteins, the histones, and quantities of acidic proteins. Each species has a characteristic number of chromosomes present in each cell. The chromosome number is maintained by mitosis.

In many eukaryotes, it is possible to distinguish between cells of the germ line which are involved in the transfer of genetic information from one generation to the next, and somatic cells which are involved in vegetative development. In the latter, quantitative changes in the DNA content of the cell are not uncommon. Increases in the ploidy of cells occur during the differentiation of plant roots, and the polytene chromosomes of the salivary glands have been extensively studied in the Dipterans, *Chironomus* and *Drosophila*. In the coccid bug *Steatococcus tuberculatus*, nuclei of the Malphigian tubules contain 32 × the haploid amount of DNA. The biochemical and cytological basis for this change has not been established. In the ciliate *Stylonichia*, a small portion of the DNA in the vegetative macronucleus is amplified to a final level of 65 × above that of the micronucleus.

Qualitative changes in the DNA content of cells during differentiation have also been reported in several species of the Diptera, where the germ line contains 4 or 6 S chromosomes and up to 52 E chromosomes. These E chromosomes do not occur in somatic tissues and are lost during the first few cleavage divisions.

In addition to chromosomal DNA, eukaryotic cells also contain cytoplasmic DNA. The existence of DNA in mitochondria and plastids is well established. DNA has also been reported to be present in centrioles, e.g., *Tetrahymena pyriformis*, in the granules (parabasal bodies) at the base of the flagella in certain protozoa, and in yolk sacs in the oocytes of sea urchins. The parabasal body clearly has a role in the morphogenesis of the flagellum and the centriole may play an important role in differentiation as it controls the orientation of the metaphase plate during cell division. The DNA in mitochondria and plastids has been shown to play an important role in controlling the morphogenesis of these organelles.

Recent studies have indicated that chromosomes of many eukaryotic cells may contain DNA derived from oncogenic viruses. This DNA may be the viral genome itself or transcribed from the genome of an RNA virus by a reverse transcriptase. Such viruses have a marked effect on the properties of the host cell. Whether these viruses can act as a model for certain developmental processes in eukaryotes as phage have done for bacterial development remains to be seen.

DNA REPLICATION

Prokaryotes. DNA reproduces itself by the process of semi-conservative replication. In this, the two DNA strands separate and new strands are synthesized opposite each of the old strands, giving rise to two double stranded molecules, each of which contains one old strand and one new one. This method of replication is dependent upon the fact that the deoxyribonucleotides always pair in the same manner; A to T and G to C. The precursor molecules: dATP, dGTP, dCTP and dTTP arrange themselves along each existing single strand of DNA which acts as a template, so that the sequence of bases is determined by this template and not by the enzyme, DNA polymerase, responsible for the formation of the phosphodiester bonds between nucleotides. Each new DNA strand is complementary rather than identical to its template strand (Fig. 1.2). Several enzymes have been described which catalyse DNA replication, both soluble and membrane bound: however, the detailed mechanism by which a circular DNA molecule containing two anti-parallel strands replicates itself has not yet been elucidated. In bacteria, DNA replication does not appear to be important in the control of cellular morphogenesis other than being an essential stage in the cell cycle.

Eukaryotes. Autoradiographic evidence suggests that DNA is replicated semi-conservatively in eukaryotic cells. Several DNA polymerases have been isolated, but it has not been shown which, if any, is concerned with chromosomal DNA replication, and which with other functions such as mitochondrial DNA replication, DNA repair, or gene amplification. After DNA replication, daughter chromosomes are distributed equally between the two daughter nuclei by mitosis such that each receives an identical set. The orientation of the mitotic spindle and the metaphase plate appears to play an important role in cell differentiation since in many plants and animals, cell wall formation is initiated along the metaphase plate of the latest nuclear division. The size of a cell is determined by the amount of growth which occurs between each cycle of cell division. In uninucleate cells, nuclear division must occur at the the same rate as cell division and must precede cell separation, hence the control of nuclear division may be an important factor in controlling

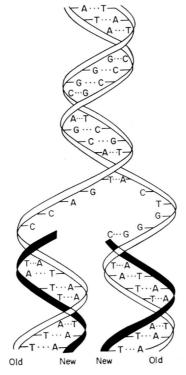

Fig. 1.2. DNA structure and replication.

cell size. In many fungi and algae, multinucleate cells can arise as a result of nuclear division occurring more rapidly than cell division. In these organisms septum formation occurs independently of the process of nuclear division.

Nuclear transplant experiments in *Acetabularia* have shown that if a nucleus is transferred from a young cell to a mature cell in which nuclear division was about to take place, then the transplanted nucleus divides. The rate of nuclear division appears to be controlled by the cytoplasm. This conclusion has been supported by similar experiments in the ciliate *Stentor* and by a somewhat different line of evidence obtained from studies on *Physarum* (Chapter 3).

INFORMATION TRANSFER AND THE GENETIC CODE

Genetic expression requires the transfer of information contained in the sequence of bases in DNA to the sequence of amino acids in protein.

Since amino acids have no obvious chemical affinity for bases, they cannot be arranged in the correct linear manner simply by using DNA as the template, as are riboncleotides during RNA synthesis. The information coded in DNA is first transcribed into a complementary strand of RNA, called messenger RNA (mRNA), which in turn acts as the template for protein synthesis. The mRNA associates with ribosomes forming a complex which is the active unit of protein synthesis.

Translation of the linear sequence of polynucleotides in mRNA into a

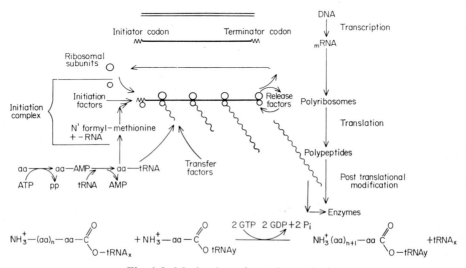

Fig. 1.3. Mechanism of protein synthesis.

linear sequence of amino acids in proteins is achieved by means of specific adaptor molecules—the transfer RNAs (tRNA). These molecules have several active sites. One of these is the anticodon, a sequence of bases complementary to the codon for one amino acid on the mRNA. Another is the amino acid binding site to which the corresponding amino acid is bound. The specificity of this reaction involves the binding of a specific enzyme at a third site. The amino acid sequence of the protein is therefore determined by the arrangement of linear tRNA molecules on the mRNA and not by the enzymes which subsequently catalyse peptide bond formation (Fig. 1.3).

RNA has a similar chemical structure to DNA except that the nucleotides contain ribose rather than deoxyribose and the base thymine is replaced by uracil. Extensive studies, mostly on bacteria and phage, have shown that the genetic code is read in groups of three bases or triplets starting at the 5-OH end of the RNA strand and running continuously along it in groups of three. Since there is a choice of four bases at each of the first, second and third positions in the triplet then there are 64 possible triplets to code for the twenty amino acids found in proteins. Using a variety of techniques, it has been shown that several different triplets can code for each amino acid and also that there are three triplets UAA, UAG and UGA which do not code for any amino acid. These three triplets control the termination of translation. There is no unique codon for chain initiation; however, the triplet AUG which codes for methionine appears to serve this function.

RNA Synthesis

Prokaryotes. Enzymes have been isolated which catalyse the polymerization of the ribonucleotides ATP, GTP, CTP and UTP into a single stranded RNA molecule using DNA as a template. The sequence of bases in RNA is complementary to the template DNA and not affected by the DNA dependant RNA polymerase. Since only one of the strands of the double stranded DNA molecule acts as a template for transcription *in vivo*, the base composition of the RNA produced will be identical to that of the non-transcribed strand with the exception of the T to U substitution.

Messenger RNA appears to be the direct product of the transcription process. Transfer RNA and ribosomal RNA precursors on the other hand are modified after transcription; for example, many bases are methylated. It is not clear whether these post-transcriptional events are important in cell regulation in prokaryotes.

Studies on enzyme induction and repression in *Escherichia coli* and other bacteria indicate that differential enzyme production is frequently the direct consequence of the differential production of the corresponding mRNAs. The induction of such an enzyme, β-galactosidase, has been studied intensively in *E. coli*. β-Galactosidase is one of three enzymes which are produced when *E. coli* is grown on a medium containing lactose as the sole carbon source. The genes for the three enzymes, β-galactosidase, galactoside permease and galactoside acetylase, are adjacent to each other on the chromosome of *E. coli*. The production of mRNA from these three genes is controlled by two other regions; the operator, which is adjacent to the β-galactosidase gene, and the regulator gene. The product of the regulator gene, the repressor protein, is produced constitutively. In the absence of lactose, the repressor protein binds to the operator and inhibits the transcription of the three adjacent genes. Since all three are controlled by the same operator, they are referred to as the "lac operon". The repressor protein is an allosteric protein which has an affinity for lactose. In the presence of lactose or other β-galactosides, it is modified allosterically and its affinity for the operator is reduced. In this way, the repression of transcription is

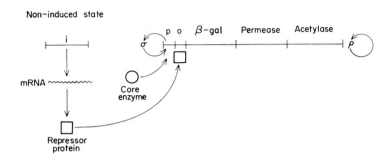

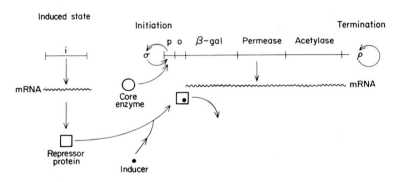

Fig. 1.4. Mechanism of enzyme induction in the "lac" operon.

relieved and mRNA synthesis of the three structural genes is initiated (Fig. 1.4).

The synthesis of the enzymes involved in histidine biosynthesis is repressed when histidine is present in the medium. Analysis of this system has shown that the regulatory mechanism is very similar to the one operating during β-galactosidase induction. The structural genes for the enzymes involved are adjacent on the chromosome and controlled by a single operator. In this system, however, the repressor protein has no affinity for the operator so transcription occurs in the absence of histidine. If histidine is supplied or accumulates within the cell, it acts on the repressor protein causing an allosteric change in conformation. The histidine-repressor complex binds to the operator and represses transcription.

In the lac operon, the DNA dependent RNA polymerase binds to a site adjacent to the operator, known as the promotor. This promotor site is involved in a second method of regulation of the β-galactosidase operon, catabolite repression. β-Galactosidase synthesis does not occur if glucose is present in the medium, in addition to lactose. Catabolite repression is a general phenomenon which affects the biosynthesis of a large number of enzymes in both prokaryotes and eukaryotes. Glucose does not act directly on the promotor but indirectly by controlling the level of cyclic AMP in the cell. In these cases, the presence of cyclic AMP appears to be essential for the recognition of the promotor site by DNA-dependent RNA polymerase. In the presence of glucose, the level of cyclic AMP is low and the polymerase fails to bind to the promotor of the lac operon and others.

In the systems referred to so far, transcription occurs when the repressor protein is absent or in its inactive form; such a system is referred to as being under negative control. Another system has been described in which the product of the regulator gene exerts a positive control on transcription. A series of enzymes involved in the conversion of arabinose to D-xylulose, 5-phosphate are induced by arabinose. The structural genes for these enzymes are controlled by a regulator gene C which differs from the regulator genes referred to above in that its product is essential for transcription.

DNA dependent RNA polymerase is a complex enzyme containing five different polypeptide chains, β', β, α, ω and σ. One of these, the σ factor, is unusual in that it is only loosely attached to the remainder of the enzyme molecule, the core enzyme, which contains $\beta'\beta\alpha_2\omega$ subunits. The σ factor is not essential for the catalytic function of the enzyme, but is concerned with the recognition of initiation sites for transcription on the DNA molecule. Once this function has been achieved, the σ factor can dissociate itself from the core enzyme, and is free to attach itself to others. Evidence from studies on phage infection and bacterial sporulation indicates that different σ factors control the transcription of different blocks of genes.

The information available at the present time thus suggests that differential enzyme synthesis can be controlled by differential mRNA production. A single mRNA may code for several proteins but each mRNA represents only a small part of the bacterial genome. Differential mRNA production requires not only specific initiation mechanisms but also specific termination mechanisms, and a protein factor ρ has been isolated which serves this function.

Eukaryotes. There is no reason to believe that the enzymic mechanism of transcription in eukaryotes is very different from that occurring in prokaryotes. Enzymes, which synthesize RNA from a mixture of ribonucleotides using double stranded DNA as the template, have been isolated from many tissues.

It is probable that the 1000-fold increase in the amount of DNA in eukaryotic cells when compared with prokaryotic cells, mainly represents an increase in the regulatory complexity of the cell. Because of this, and also because such detailed genetic analysis has not been carried out in most eukaryotic cells, our understanding of the regulatory systems operating in them is limited. The inhibition of development by actinomycin D, an antibiotic which inhibits DNA and RNA synthesis, has been taken to indicate a requirement for the synthesis of new mRNAs and new enzymes for the process. However, such evidence is at best circumstantial. More direct evidence has been obtained using DNA : RNA hybridization techniques.

Since all RNA has a base sequence which is complementary to that of one of the two DNA strands, it can be made to bind to this DNA in a specific manner. The experiments are usually carried out using isotopically labelled RNA, so that the amount of RNA bound can be measured accurately. Since the number of sites on a given amount of DNA which can bind one species of RNA are limited, it has been possible to estimate by competition experiments whether two RNA fractions are the same or different. If the DNA preparation is incubated with an excess of one type of RNA, then all the binding sites will be occupied and no further binding will occur if a second sample of identical but isotopically labelled RNA is added. If the second species of RNA is not identical, then binding of this may occur at a different site. Using various modifications of this technique, it has been shown that different RNA fractions are produced in different mammalian tissues.

Histones, the basic proteins found in eukaryotic chromosomes, may appear to be the natural candidates for the repressors of gene action in eukaryotic cells. Unfortunately, chemical studies indicate that they do not possess sufficient molecular diversity to function alone as regulator proteins. They may, however, play an important role in conjunction with other more specific molecules.

Using the DNA : RNA hybridization technique, it has been shown that a majority of the DNA in chromosomes is masked and cannot bind added RNA. The unmasked segments of DNA differ in chromatin isolated from different tissues. Chromatin has been studied in an attempt to establish which proteins are responsible for DNA masking. The results indicate that histones are active in this respect, but also that acidic proteins which occur in much smaller quantities also have an influence. Histones can be modified by phosphorylation, acetylation or reduction *in vivo*. In regenerating rat liver, acetylation of histones is highest when RNA synthesis is highest. Similar correlations between the degree of chemical modification of the histone fraction and the rate of RNA synthesis have been observed in other systems. In *in vitro* systems, histones have been shown to inhibit transcription. The degree of inhibition varies with different histone fractions. Phosphorylated, acetylated and reduced fractions, in particular, are less effective as inhibitors of DNA dependent RNA synthesis.

Other evidence for the differential production of RNA during eukaryotic differentiation has been obtained from studies on the polytene chromosomes of *Drosophila* and *Chironomus*. It has been shown that specific regions of these chromosomes become disorganized at different stages of development, forming structures referred to as puffs. Recent autoradiographic studies show that these puffs are regions of intensive RNA synthesis. In many organisms which do not have polytene chromosomes, different chromosomal regions exhibit different staining reaction with Feulgen; the euchromatic regions which appear diffuse, and the heterochromatic regions which remain compact, densely stain during interphase. It is thought that the euchromatic regions are regions of high RNA synthesizing activity and the heterochromatic regions are those of low activity. In developing lymphocytes, it has been demonstrated that labelled acetate is incorporated into heterochrometin more rapidly than euchromatin. Euchromatic and heterochromatic regions of the chromosomes often vary throughout the cell cycle and life cycle.

In eukaryotic chromosomes in general therefore, it appears that the template activity of large segments of the chromosome can be suppressed. In female mammals, the whole of one X chromosome remains non-functional throughout the cell cycle. This inhibition of activity is not, however, the result of a permanent change in the chromosome, since in different cells the identity of the functional and non-functional X chromosome appears to be determined randomly.

Three lines of evidence suggest that gene amplification may be an important mechanism in the control of RNA synthesis:

(1) During puff formation, rapid DNA synthesis occurs and the level of DNA increases to a level far higher than before puffing.

(2) An analogous phenomenon to that of puffing in polytene chromosomes may occur in certain amphibia. The chromosomes of newt oöcytes exhibit an unusual morphology and are referred to as

lampbrush chromosomes. They consist of a central backbone of compact DNA with large irregular loops of DNA of between 10 and 100 μm in length situated along it. These must contain sufficient DNA to code for either several different proteins or many copies of the same protein. Each of these loops produces a single mRNA. Whether or not these loops consist of amplified DNA is not clear.

(3) The gene for rRNA has been shown to be associated with the nucleolus in eukaryotic cells. In most organisms, only one nucleolus is present per haploid number of chromosomes, but in amphibian oocytes up to 1000 nucleoli may be present. Each of these has been shown to contain many copies of the gene for rRNA, arranged in tandem on a single DNA molecule.

In eukaryotes, it has been established that the nucleus contains a large amount of high molecular weight RNA known as heterogeneous nuclear RNA. Recent studies indicate that part of this may be mRNA precursor, but large amounts are broken down, probably in the cytoplasm. The production of rRNA from the rDNA gene is a better documented example of the relation between the immediate and the final gene product. The immediate product which sediments at 45 S is a precursor for the RNA present in both ribosomal subunits. These subunits are produced from this precursor by a series of cleavages which result in almost half the RNA present being discarded (Fig. 1.5).

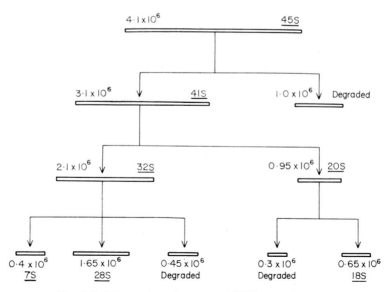

Fig. 1.5. Maturation of ribosomal RNA in eukaryotes.

In general, little is known of the mechanism of enzyme induction and repression in eukaryotic cells. Evidence from both animals and plants indicates that hormones can induce changes in the enzyme composition of cells and in some cases, increased RNA synthesis has been observed. It seems unlikely that hormones act directly at the chromosomal level but they probably interact with other molecules which transfer the hormonal "message" to the chromosome. Such molecules have not been identified, but it seems probable that in some cases, cyclic AMP is involved in this process. Hormones provide a mechanism by which different cells, tissues or organs in eukaryotic organisms can influence each other's metabolism and development. In higher animals, hormones which act between different individuals are referred to as pheromones. It is interesting to note that in micro-organisms this distinction has not been made.

PROTEIN SYNTHESIS

Prokaryotes. The process of polypeptide synthesis on the ribosome-mRNA complex is referred to as translation, since it is at this level that the linear sequence of bases present in mRNA is translated into a linear sequence of amino acids. The tRNA molecule is the critical link in this process. Transfer RNA molecules, which can recognize the various codons, occur in the cell. Each tRNA molecule contains a sequence of three bases, "the anticodon", which is complementary to one or more codons. The tRNA molecules cannot recognize amino acids directly, but amino acid activating enzymes exist which recognize individual amino acids and their corresponding tRNA molecules. These enzymes catalyse the formation of an ester bond between the carbonyl group of the amino acid and the 3'-OH group of adenine which occurs at the 3' end of all tRNA molecules (Fig. 1.3).

The ribosome appears to provide the correct molecular environment for the interaction of mRNA, the amino acid tRNAs and various enzymes and co-factors, thus permitting peptide bond formation. The complex events occurring in this process are summarized in Fig. 1.3.

Translation is initiated by the formation of an initiation complex between N'-formyl-methionine tRNA, mRNA, the 30 S ribosomal subunit and various initiation factors. The 50 S ribosomal subunit then becomes attached to this complex. The peptide chain is formed as the ribosome moves along the mRNA and is terminated when the ribosome reaches the chain termination codon in the mRNA. There is no tRNA molecule corresponding to the chain termination codons, which appear to be recognized by specific protein factors called release factors.

Although N'-formyl-methionine is the first amino acid incorporated into all polypeptide chains, it rarely appears in the mature proteins. In some instances, the formyl group is removed, leaving methionine as the N'-terminal amino acid while in others, methionine itself is removed. In active protein synthesizing systems it is normal to find several ribosomes complexed with each mRNA molecule, giving rise to a polysome.

Not all the proteins specified by a single mRNA are synthesized at the same rate. In the β-galactosidase operon, the numbers of molecules of β-galactosidase, permease and acetylase are produced in a ratio of $1 : 0.5 : 0.2$. Some mechanism preventing the translation of the whole mRNA by at least some ribosomes must therefore exist. Several hypotheses have been presented to explain this type of translational control.

(1) Some ribosomes may "drop off" or be removed from the messenger at the end of the galactosidase or permease gene.

(2) The rate of transcription of those enzymes which are produced in smaller amounts may be slower. It has been suggested that some tRNA molecules are in short supply and could limit the rate of transcription. Such a mechanism would result in an accumulation of ribosomes at a point proximal to the triplet for that tRNA.

(3) Ribosomes can sometimes attach at the initiation regions of genes in the middle of a mRNA molecule. In this case, there may be differential affinity of these initiation regions for the ribosome. This may be attributable to specific base sequences or to differences in the tertiary structure of the mRNA.

(4) Initiation sites may be blocked by specific proteins which inhibit the binding of ribosomes.

Base sequencing of the RNA phages R_{17} and $Q\beta$ have shown that termination codons do not occur here singly but in twos or threes at the end of a gene. This may indicate that the number of termination codons may control the rate of ribosome "drop off". If this is so, then regulation of translation could be controlled by the number of termination codons interspersed between different genes on the same mRNA.

No direct evidence is available for the hypothesis that a low frequency of a tRNA can restrict the rate of translation of certain proteins. However, it has been observed that different tRNAs coding for the same amino acid are present in different amounts and that their relative abundance varies at different stages of development, e.g. bacterial sporulation.

In phage R_{17} and its relatives three proteins, the attachment protein (A), the coat protein (CP) and an RNA synthetase (SYN) are produced in the ratio of $1 : 10 : 1$. Since the genes occur in the sequence A, CP,

SYN on the RNA molecule, then regulation at the level of translation must occur. It appears that each gene has an initiation site and that the initiation site of CP has a much higher affinity for ribosomes than the other two.

Ribosomes are prevented from moving along the mRNA from the CP to the SYN gene by coat protein which binds to the initiation site of the SYN gene. Thus the product of the CP gene acts both as a structural and as a regulator protein.

Eukaryotes. The translation process in eukaryotes differs at least in detail from that occurring in prokaryotes. The most conspicuous difference is that eukaryotic ribosomes sediment at 80 S, compared with the 70 S prokaryotic ribosomes, and contain subunits which sediment at 60 S and 40 S. Each subunit contains more RNA and more protein than its prokaryotic counterpart. Functional differences are apparent in the differential sensitivity of translation in eukaryotes and prokaryotes to certain antibiotics. Chloramphenicol inhibits translation in bacteria but not eukaryotes, while the reverse is true for cycloheximide. Protein synthesis in plastids and mitochondria resembles the prokaryotic system in its general properties.

Translational control is difficult to demonstrate in eukaryotes, since genetic and biochemical studies are not sufficiently advanced in most cases to indicate whether two or more proteins are produced by the same mRNA. However, observations have been made which suggest that translational control may occur. Changes in the levels of isoaccepting species of tRNA have been observed at different stages in the development of eukaryotic cells and may be involved in translational control.

More direct evidence is available for the existence of stable mRNA. It is apparent that the amount of protein synthesized from a given mRNA can be affected by its stability. In rapidly growing HeLa cells, the average life of an mRNA molecule is three hours. In erythrocytes, on the other hand, no RNA synthesis occurs after the onset of haemoglobin synthesis so existing mRNA molecules appear to be stable for up to 100 h. Reticulocytes are abnormal in that the nucleus is lost at an early stage in development. However, mRNA of similar stability has also been reported in liver cells which are responsible for the synthesis of plasma proteins.

Stable mRNA which appears to remain in a non-functional state for long periods of time has been observed in the eggs of echinoderms, amphibians and other animals. In amphibians, development can continue to the end of the blastula stage in the presence of concentrations of actinomycin D, which inhibit RNA synthesis. Addition of puromycin, an inhibitor of translation, rapidly terminates development. Development has been shown to be dependent upon the existence of masked mRNA in the cytoplasm. This masked mRNA appears to be in the form of a stable, polysome complex which is not active in protein synthesis and is resistant to RNAase. This complex can be activated by treatment with trypsin, suggesting that the stability can be attributed to a protein inhibitor present in the complex. Evidence indicating that different masked mRNAs are activated at different times during blastulation has been presented.

Polyribosomes in eukaryotic cells are frequently associated with the membranes of the endoplasmic reticulum. Although there is still some controversy as to whether membrane bound or free polyribosomes are most active in protein synthesis, it now seems likely that both are important in most cells. Studies using the electron microscope indicate that membrane bound polyribosomes are most abundant in cells which secrete large amounts of protein, e.g. gland cells and the cells in rat liver producing plasma proteins. Many workers consider that the membrane in these cells is primarily concerned with secretion of the protein, and not with its synthesis. Membrane-bound polyribosomes can be readily released by treatment with the detergent sodium deoxycholate (DOC) and are capable of carrying out protein synthesis in the absence of the membrane fraction. In rat liver cells, it has been shown that labelled amino acids first appear in serum albumin associated with the microsome fraction. Similar results have been obtained in studies on amylase synthesis in pancreatic cells from pigeons. In these experiments, the amylase and serum albumin were liberated from the microsomes by treatment with DOC. It seems possible that membranes may be important in controlling the biosynthesis of extracellular enzymes and enzymes which are localized within vesicles.

Several hypotheses have been put forward suggesting that post-translational modification of polypeptides may be important in the control of enzyme synthesis. It is possible that globin release from polyribosomes may be influenced by the haem prosthetic group. Many other enzymes containing prosthetic groups may be controlled in a similar manner.

Fungal Studies, the State of Play

By reference to work carried out on the bacteria and higher eukaryotes, an attempt has been made to indicate the complexity of the different biochemical processes which may be involved in any one developmental sequence. Throughout this book, it will be apparent that the fungi have been less intensively studied than certain other groups, and that in general only a few of the many techniques available have been applied to any one system. In this final section of the introduction, the results which indicate that the basic cellular processes of fungi resemble those of other eukaryotes are summarized. Much of the data referred to is covered in greater detail in subsequent chapters.

The fungi appear to be one of the simplest groups of eukaryotes. The quantity of DNA per cell varies from four to ten times the amount present in a cell of *Escherichia coli*. This figure can be compared with the 1000–10,000 fold increase observed in the higher plants and animals. In the fungi, the majority of the DNA is found within the well-defined nucleus in which it is aggregated into distinct chromosomes which have also been shown to contain histones (Duffus, 1971). In addition, DNA is present in mitochondria, and in certain species, e.g. *Allomyces* and *Blastocladiella*, a DNA particle has been described which appears to play a role in morphogenesis (Chapter 4). However, not all cytoplasmically inherited characteristics have been associated with cytoplasmic DNA. The discovery of viruses in fungi (Banks *et al.*, 1969) and a double stranded RNA particle in yeast which controls the killer phenotype (Berry and Bevan, 1972) opens up the possibility that such particles may be responsible for other cytoplasmically inherited characteristics.

Little information is available on the enzymology of DNA replication in the fungi. In *Physarum polycephalum*, it has been reported that the level of a soluble DNA polymerase was identical in the S and G_2 phases of the cell cycle suggesting that if this is the enzyme involved in DNA synthesis then the level of the enzyme is not the critical controlling factor (Sauer, 1973). In yeast, two enzymes have been described but the precise role of each has not been established.

The mechanics of nuclear division are not as easily studied in the fungi as in other groups since the chromosomes are small and not clearly defined at mitosis. However, it appears that nuclear division has the basic characteristics of a mitotic division although considerable variation between species has been observed (Burnett, 1968). The nuclear membrane remains intact throughout nuclear division. The relationship between nuclear division and cell division has been studied intensively in *Physarum polycephalum* and *Schizosaccharomyces pombe* (Chapters 3 and 5). The vegetative cell cycle is being revealed as a delicately balanced system involving the control of transcription as well as DNA replication and cell division. In *Schizophyllum commune* and other basidiomycetes, the site of cell wall formation is determined by the site of nuclear division which in turn is controlled in such a way that the dividing nucleus is always to be found at a point half-way between the hyphal apex and the first septum. Since in *S. commune* branching always occurs at the site of the clamp connection, the regulation of the cell cycle can be clearly seen to play a major role in controlling the morphology of the dikaryotic mycelium (Chapter 7). In multinucleate cells of *Neurospora crassa*, septum formation is not associated directly with mitosis; however, it appears that the DNA : cytoplasm ratio is constant since diploid strains have half as many nuclei per cell as do haploid strains (Clutterbuck, 1969).

The mechanisms of RNA and protein synthesis in fungi appear to be similar to those described for other eukaryotic cells. Large (24 S) and small (17 S) ribosomal subunits, transfer RNA, and fractions which have the characteristics of messenger RNA have been isolated from yeast and other fungi (Mounolou, 1971). The yeasts in particular have been used for many fundamental studies in cell biology. Yeast transfer RNA was used by Holley (1963) for pioneering studies on the structure of transfer RNA and many of the studies describing an autonomous protein synthesizing system in mitochondria have been carried out on yeast. Recent studies on the development of mitochondria when anaerobically grown yeast is transferred to aerobic conditions indicate that yeast provides an ideal system for studying the biogenesis of mitochondria (Linnane *et al.*, 1972).

Although a DNA dependent RNA polymerase has been partially purified from yeast and three different enzymes have been reported in *Rhizopus stolonifera* (Chapter 4) details of the mechanism of RNA synthesis in fungi remain in doubt. As yet no enzyme component which is equivalent to the σ factor isolated from bacterial systems has been identified. It has been shown that several copies of rRNA cistrons are present in the yeast genome (Chapter 5) and that rRNA is initially transcribed as a 38 S precursor molecule (Robichon Szulmajster *et al.*, 1971). The mechanism of messenger RNA formation remains an enigma. Studies on synchronously growing cells of *Schizosaccharomyces pombe* indicate that both constitutive and induced enzymes are produced at a specific point in the cell cycle (Chapter 5). These results have been interpreted as indicating that transcription proceeds from one end of a chromosome to the other during the cell cycle. Both enzyme induction and repression have been described in the fungi. However, the evidence that these processes involve differential mRNA synthesis is far less convincing than that presented from bacterial studies. Differential transcription has been demonstrated during sporulation in *Physarum polycephalum* using a hybridization technique, and a requirement for RNA synthesis for the production of enzymes required during different developmental sequences has been observed in several species (Chapters 2, 3 and 5).

There has been considerable disagreement as to the value of studies in which actinomycin D, an inhibitor of DNA dependent RNA synthesis, has been used to demonstrate a requirement for RNA synthesis. It has been pointed out by Wright (1972) that actinomycin D, in addition to inhibiting transcription, also inhibits respiration, glycolysis, protein synthesis, phospholipid synthesis and amino acid transport. It has also been shown to interfere with the assembly of the Golgi apparatus and to stimulate mRNA breakdown. Thus, for any particular system studied, it is difficult to be confident that the inhibition of transcription is the primary action of the drug. Wright also emphasizes that it should be established that changes in the activity of an enzyme can be attributed to changes in enzyme concentration rather than changes in the levels of substrates, cofactors, activators and inhibitors or other changes in the cellular environment of the enzyme. The use of actinomycin D in studies on *Dictyostelium discoideum* has been defended by Sussman (1972). The level of actinomycin D required to inhibit enzyme formation during sorocarp formation was shown to be similar to that required to inhibit RNA synthesis in isolated nuclei. Furthermore, the period of time required for actinomycin D to act varied with the physiological state of the cell, suggesting that it had no effect until further messenger RNA synthesis was essential for development. That an increase in enzyme activity was associated with *de novo* synthesis of the enzyme was clearly demonstrated in the case of UDPG pyrophosphorylase. In these experiments newly synthesized enzyme was labelled by feeding the organism ^{35}S methionine, ^{3}H leucine or ^{14}C amino acids, then isolated using a combination of immunological and electrophoretic techniques. It is evident that although actinomycin D is a valuable inhibitor for developmental studies, it is desirable where possible to carry out further experimental studies to confirm the mode of action of the drug.

The role of transcription in fungal differentiation deserves further study since in several fungi it has been observed that complex changes in structure and morphology can occur in the absence of RNA synthesis, e.g. planospore germination (see Chapter 4).

The inhibition of a developmental process by high concentrations of glucose is a common phenomenon in fungi (Chapter 8). It is not clear, however, how closely these phenomena can be compared to catabolite repression in bacteria. It has recently been reported that the inhibition of sporulation in yeast by glucose can be reversed by the addition of cyclic AMP which points to a similarity of mechanism (Chapter 7).

Although detailed studies on the mechanism of translation have not been carried out in fungi, with the possible exception of yeast (Robichon-Szulmajster *et al.*, 1971), results obtained from *in vivo* studies on poly-ribosome formation and *in vitro* studies using polyuridylic acid as a messenger indicate that the mechanism is again similar to that des-cribed for other eukaryotes (see Chapters 2, 3 and 4). *De novo* protein synthesis has been demonstrated using both density-labelling and anti-genic techniques.

Changes in the isoenzyme composition have been observed in several differentiating systems, e.g. dikaryon formation in *Schizophyllum commune* (Chapter 7) and spherule formation in *Physarum polycephalum* (Chapter 3) but it has not been established whether the observed changes are the result of the synthesis of new enzymes or attributable to changes in the conformation of existing enzymes.

Most of the material referred to above is concerned with the control of the timing of enzyme synthesis and the amount of enzyme produced. Other important control mechanisms are those concerned with the distribution of biosynthetic products within the cell. Several studies referred to in subsequent chapters suggest that vesicles play a critical role in transporting enzymes to specific sites in the cell and that many structural cell components are synthesized *in situ* (Chapter 5). In parti-cular, the elegant work of Bartnicki-Garcia illustrates how deposition of wall material in the apical region of a hypha can give rise to tubular growth. The mechanism by which this polarity of growth is established during spore germination remains an intriguing mystery.

The fungi also provide valuable experimental material for studies on organelle mobility. The rapid movement of nuclei through the mycelium of Basidiomycetes prior to dikaryon formation is a remarkable pheno-menon, and the observation that no general cytoplasmic exchange occurs during this process poses interesting problems as to the mechan-ism of nuclear movement (Chapter 7).

Although fungi possess many of the advantages of unicellular organ-isms for experimental analysis, they also form multicellular structures and so can provide model systems for the study of the organization of multicellular organisms. The aggregation of the amoebae of *Dictyo-stelium discoideum* and the subsequent formation of the sorocarp have always fascinated cell biologists. The extensive studies on this system which have indicated the importance of cyclic AMP in cell aggregation, the importance of specific cell–cell interactions in establishing the multicellular state and the existence of precise mechanisms of cellular assortment during sorocarp formation have made a significant con-tribution to our understanding of the multicellular state. The factors controlling the development of fruit-bodies in the higher fungi are not well understood, but it is difficult to see how such complex structures could be controlled without the existence of intercellular messages. However, such substances have not been identified with confidence (Chapter 7). The control of branching is discussed in Chapter 8 in the context of growth rates. However this, together with the control of anastomosis, must play an important role in controlling the development of multicellular structures in the higher fungi.

The fungi are an important group of organisms which deserve study in their own right. Since they also represent one of the simplest groups of eukaryotic organisms, it can be anticipated that studies on the fungi should contribute to our understanding of eukaryotic biology as a whole. It would be no exaggeration to say that this has already occurred in the case of the studies on *Dictyostelium* and *Physarum* and it seems probable that, as a result of the increasing interest in fungal development, further advances will be made.

"There are many reasons to work on organisms like yeast."
(Watson, 1970)

Summary

An understanding of morphogenesis requires not only a knowledge of the type of molecules present in a cell and how they are synthesized but also an understanding of the mechanism by which they are arranged in space. Since most fungi exhibit a range of form, the study of morphogenesis must involve a study of the mechanism of differentiation. An understanding of this involves an integration of physiological, biochemical, cytological and genetic information relevant to the cell structures being studied. There are probably few studies which are not relevant to differentiation. In our view, however, the genetic and biochemical mechanisms which are directly involved in the regulation of differential gene expression and the biosynthesis of informational macromolecules are central to the problem of differentiation. Many of the mechanisms which have been observed in the bacteria and higher mammals have not yet been described in the fungi. However, such information as is available indicates that similar mechanisms do operate in the fungi.

One advantage of studying fungi rather than bacteria is that they offer an opportunity to study the mechanisms by which cells interact to produce a multicellular organism.

Recommended Literature

Concepts of Morphogenesis and Differentiation

Bonner, J. T. (1973). Development in lower organisms. *Symposium Society General Microbiology* **23,** 1–7. Cambridge University Press, London.

Gross, P. R., (1968). Biochemistry of differentiation. *Journal of American Chemical Society* **75,** 631–660.

Pasternak, C. A. (1970). "Biochemistry of Differentiation". Wiley Interscience, London.

Schjeide, O. A. and De Vellis, J. (1970). Ed. "Cell Differentiation". Van Nostrand, Reinhold Company, New York.

Waddington, C. H. (1966). "Cellular Differentiation". Prentice Hall International, Hemel Hempstead.

Wright, B. E. (1964). The biochemistry of morphogenesis. *In* "Comparative Biochemistry". Ed. Florkin, M. and Mason, H. S. Academic Press, London and New York.

Regulation of Enzyme Activity

Cohen, G. N. (1968). "The Regulation of Cell Metabolism". Holt, Rinehart and Winston, Inc., New York.

Gustafson, G. L. and Wright, B. E. (1972). Analysis of approaches used in studying differentiation of the cellular slime moulds. *Critical Reviews In Microbiology* **I,** 453–478.

De Vellis, J. (1970). Enzyme regulation during cell differentiation. *In* "Cell Differentiation". Ed. Schjeide, O. A. and De Vellis, J. Van Nostrand, Reinhold Company, New York.

General Reviews on Cell Biology

Charles, H. P. and Knight, B. C. J. G. (1970). Ed. "Organisation and Control in Prokaryotic and Eukaryotic Cells. *Symposium Society General Microbiology* **20.** Cambridge University Press, London.

Davidson, J. N. (1969). "Biochemistry of Nucleic Acids". 6th Edn. Methuen, London.

Hartman, P. E. and Suskind, S. R. (1969). "Gene Action". 2nd Edn. Prentice Hall, New York.

Ingram, V. M. (1972). "Biosynthesis of Macromolecules". 2nd Edn. W. A. Benjamin, Inc., New York.

Watson, J. D., 1970. "Molecular Biology of the Gene". 2nd Edn. W. A. Benjamin, Inc., New York.

DNA and Chromosomes

Crick, F. (1971). General model for the chromosomes of higher organisms. *Nature* **234,** 25–27.

Dupraw, E. J. (1970). "DNA and Chromosomes". Holt, Rinehart and Winston, Inc. New York.

Lewis, K. R. and John, B. (1963). "Chromosome Marker". J. and A. Churchill Ltd., London.

Stellwagen, R. H. and Cole, R. D. (1969). Chromosomal proteins. *Annual Review of Biochemistry* **38,** 951–990.

Control of Transcription

iosick, R. and Sonensheim, A. L. (1969). Changes in the template specificity of RNA polymerase during sporulation. *Nature* **224**, 35–37.

Ptashne, M. and Gilbert, W. G. (1970). Genetic repressors. *Scientific American* **222**, 36.

Travers, A. (1971). Control of transcription in bacteria. *Nature New Biology* **229**, 69–74.

Control of Translation

Lengyel, D. and Soll, D. (1969). Mechanism of protein synthesis. *Bacteriological Reviews* **33**, 264–301.

Sueoka, N. and Kano-Sueoka, T. (1970). Transfer RNA and cell differentiation. *Progress in Nucleic Acid Research and Molecular Biology* **10**, 23–55.

Tompkins, G. M., Gelehrter, T. D., Grannder, D., Martin, D., Samuels, H. and Thompson, E. (1969). Control of specific gene expression in higher organisms. *Science* **166**, 1474–1480.

Hormones

Galston, A. W., and Davies, P. J. (1969). Hormonal regulation in higher plants. *Science* **163**, 1288–1297.

Grant, J. K. (1969). Action of steroid hormones at cellular and molecular levels. *In* "Essays in Biochemistry". Ed. Campbell, P. N. and Greville, G. D. **5**, 1–58. Academic Press, London and New York.

Tata, J. R. (1968). Hormonal regulation of growth and protein synthesis. *Nature* **219**, 331–337.

Fungal Studies

Banks, G. T., Buck, K. W., Chain, E. B., Darbyshire, J. E. and Himmelweit, F. (1969). Virus-like particles in penicillin producing strains of *Penicillium chrysogenum*. *Nature* **222**, 89–90.

Berry, E. A. and Bevan, E. A. (1972). A new species of double stranded RNA from yeast. *Nature* **239**, 279–280.

Burnett, J. H. (1968). "Fundamentals of Mycology". Chapter 13. Arnold Press, London.

Clutterbuck, A. J. (1969). Cell volume per nucleus in haploid and diploid strains of *Aspergillus nidulans*. *Journal of General Microbiology* **55**, 291–299.

Duffus, J. H. (1971) Isolation and properties of nucleohistone from the fission yeast *Schizosaccharomyces pombe*. *Biochimica et Biophysica Acta* **228**, 617–635.

Holley, R. W., Apgar, J., Everett, G. A., Madison, J. T., Merrill, S. H. and Zamir, A. (1963). *Cold Spring Harbour Symposium for Quantitative Biology* **28**, 117–121.

Linnane, A. W., Hasiam, J. M., Lukins, H. B. and Nagley, P. (1972). The biogenesis of mitochondria in micro-organisms. *Annual Review of Microbiology* **26**, 163–198.

Monolou, J. C. (1971). Yeast nucleic acid. *In* "The Yeasts". **2**, 309–333. Ed. Rose, A. H. and Harrison, J. S., Academic Press, London and New York.

Robichon-Szulmajster, H. de and Surdin-Kerjan, Y. (1971). Nucleic acid and protein synthesis in yeast. *In* "The Yeasts". **2**, 336–418. Ed. Rose, A. H., Harrison, J. S., Academic Press, London and New York.

Sauer, H. W. (1973) Differentiation in *Physarum polycephalum In* "Microbial Differentiation" **23,** *Symposium of the Society for General Microbiology*, 375–405. Ed. Ashworth, J. M. and Smith, J. E. Cambridge University Press.

Sussman, M. (1972) Letter to *Developmental Biology* **28,** f. 16–20.

Wright, B. (1972). Actinomycin D and genetic transcription during differentiation. *Developmental Biology*, **28,** f. 13–16.

2 A Cellular Slime Mould: *Dictyostelium discoideum*

2.1 Introduction

The slime moulds are a group of organisms rarely seen in nature because of their microscopic vegetative phase and fragile, minute and ephemeral reproductive structures. These organisms were first described in 1869 by a German mycologist, Oskar Brefeld, and although at least 20 species are now identified most interest has been centred around the extensive studies carried out on *Dictyostelium discoideum*. Whether or not these organisms are true fungi is debatable, though traditionally their study has very largely been the domain of the mycologist. It cannot be disputed that the studies on the physiology and biochemistry of cellular slime mould development, carried out during the last two decades, represents one of the most exciting and scientifically rewarding developments in contemporary eukaryotic cell biology. Not only have these studies set out valuable information concerning the initiation and control of the process of differentiation in simple organisms but they have also contributed significantly to a clearer understanding of the evolution of micro-organisms (Wright, 1967).

The life cycle of *Dictyostelium discoideum* is remarkable in that it is biphasic (Fig. 2.1). In the vegetative condition the organism exists as unicellular myxamoebae similar in appearance and size to human haemocytes. It inhabits damp environments engulfing bacteria which it finds by chemotaxis. Multiplication is by binary fission and provided suitable environmental conditions for growth and bacteria are present the vegetative phase will continue indefinitely. However, when the food source becomes exhausted the myxamoebae enter a period of starvation or interphase which heralds a remarkable change in development. Whereas during vegetative growth the myxamoebae were indifferent to each other they now begin to show degrees of mutual attraction leading to the formation of a well-organized community in

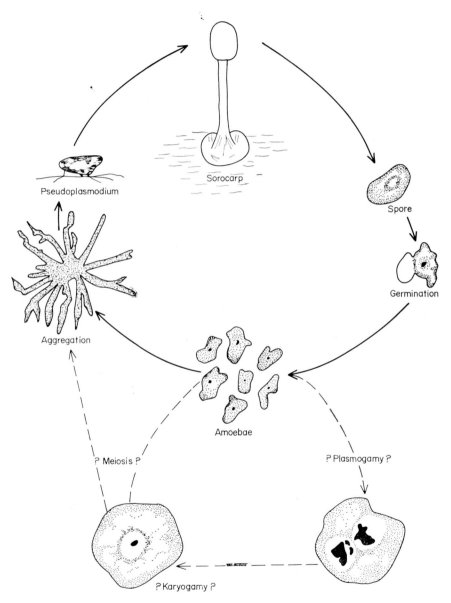

Fig. 2.1. The life-cycle of *Dictyostelium discoideum*. From Alexopoulos (1962).

which the myxamoebae, although retaining their individuality, function as one organism. During interphase, which may last between 6 and 8 h, the cell membrane undergoes antigenic and metabolic alterations. Following interphase, groups or centres of myxamoebae appear to secrete into the liquid phase of the environment a chemotactic substance which induces neighbouring myxamoebae to migrate in streams towards the initiating centres. Each aggregate, containing up to 100,000 myxamoebae, then forms a multicellular, tissue-like, slug-shaped mass, the pseudoplasmodium one or two millimetres long which is motile and can move along temperature, light and humidity gradients. Migration rates can vary between 0·3 and 2·0 mm per hour at 20°C. There is no multiplication of the myxamoebae after the onset of starvation. However, even up to late stages of pseudoplasmodium formation the process can be put into reverse by the reintroduction of bacteria. The aggregated myxamoebae will de-aggregate and resume vegetative hunting. Beyond a critical time period of pseudoplasmodium development the addition of bacteria cannot reverse the developmental programme. Thus a "point of no return" in the sequence of events that control development has been achieved; a phenomenon which will be seen repeatedly with other organisms during processes of development.

The migrating slug is enveloped in a cellulose sheath which is secreted by the anterior tip myxamoebae. This liquid sheath hardens and encases the aggregated myxamoebae which in turn move through it. Since the sheath becomes fixed to the solid phase of the medium it is left behind as a collapsed tube after the cells have passed by. This sheath has been implicated as a possible means of transmitting information from anterior to posterior cells (Ashworth, 1971).

The migrating slug is polarized and can be shown to contain two cell types, separable by isopycnic gradient centrifugation, which are biochemically and cytologically distinct. After a variable period of movement the anterior myxamoebae of the slug will upend themselves and form an erect cellulose-ensheathed stalk. The stalk myxamoebae swell and become hollow and form a slender, tapering hollow tube a few millimetres in length. As the stalk develops it carries with it the posterior myxamoebae of the slug which become a globose structure containing the spores. Approximately one-third of the slug myxamoebae will form the stalk and the remainder spores. The size of the fruiting-body can vary within wide limits and is controlled by factors such as cell density and territory size. The entire asexual reproductive structure or fruiting-body is called the sorocarp.

Thus unlike other multicellular eukaryotes *Dictyostelium discoideum* undergoes a transition from the unicellular, free-living condition found in many micro-organisms to the multicellular integrated condition associated with higher organisms. The slime moulds thus offer a unique opportunity to determine the factors that permit independent cells to adapt to a multicellular existence and the subsequent interactions that affect their course of development.

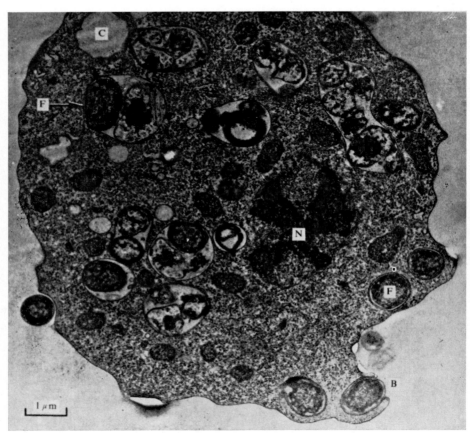

Fig. 2.2. Vegetative myxamoeba feeding on *Escherichia coli* (B) with food vacuoles (F). contractile vacuole (C), and nucleolus (N). From George, Hohl and Raper (1972).

During the various stages of differentiation of *Dictyostelium discoideum* many ultrastructural changes occur in the myxamoebae (Figs. 2.2, 2.3). At interphase food vacuoles disappear and are replaced, after aggregation, by vacuoles. During culmination the prespore myxamobae decrease in size and the cytoplasm becomes dense while the myxamoebae that will later become the central part of the stalk swell and exhibit cytoplasmic necrosis. In prespore myxamoebae profound changes occur in mitochondrial and nuclear ultrastructure. The chromatin becomes diffuse in myxamoebae of the migrating slug and by early culmination only nuclear material is identifiable. The location

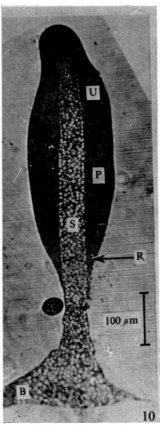

Fig. 2.3. Light micrograph of longitudinal section of sorocarp stained with Toluidine blue. Basal disc (B); dark-staining prespore cells (P); rearguard cells (R); stalk (S); prestalk cells (U). From George, Hohl and Raper (1972).

of the nuclear material is indistinct in mature spores. The mitochondria degenerate in the slug stage but reappear at early culmination in a form lacking outer membranes. The reorganization of the mitochondria is paralleled by changes in mitochondrial enzymes.

The spores of *Dictyostelium discoideum* are uninucleate with seven chromosomes and are encapsulated in a hard cellulose spore case. Until recently there was considerable controversy concerning the presence of a sexual cycle in this organism, but it has now been clearly demonstrated that as in many other lower fungi genetic recombination is by way of the parasexual cycle (Fig. 2.4). Ploidy is an inherited condition and three

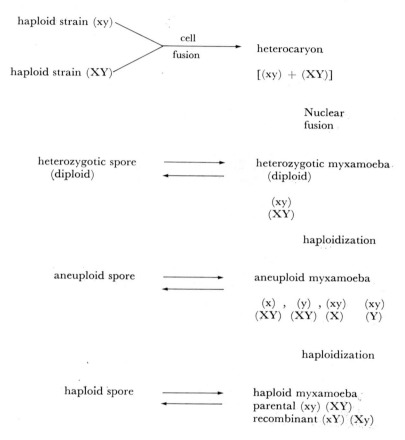

Fig. 2.4. The parasexual cycle in *Dictyostelium discoideum*. x/X and y/Y represent two unlinked genes. The genotype of the various kinds of nuclei are enclosed in parentheses. From Sinha and Ashworth (1969).

strains have been isolated: stable haploid, stable diploid and metastable. The metastable clones contain both haploid and diploid cells.

By treating myxamoebae with mutagens it has been possible to isolate strains with heritable aberrations in their developmental sequences. Such mutants either stop at a stage prior to culmination or mature with a reproductive system of abnormal character. These mutants also display changes in the normal metabolic pattern consistent with the stages at which the particular morphological aberration occurs. Mutants of *Dictyostelium discoideum* having a temperature-sensitive step either in growth processes or in the development processes of aggregation and differentiation have been isolated while mutants which affect growth, but not development, have also been shown (Sussman and Sussman, 1969).

2.2 Cultural Techniques

To produce large numbers of myxamoebae for experimental purposes *Dictyostelium* is initially grown in two-membered liquid culture with the bacteria *Escherichia coli* or *Aerobacter aerogenes*. Axenic growth of the myxamoebae has been obtained in complex rich media. The use of bacteria or chemically complex media is a serious impediment to a rational understanding of the sequential biochemical changes that occur during development. Thus with chemically defined media it is possible to vary the medium components and to ascertain the true causal relationship between internal biochemical changes and development.

Since the aggregation phase does not occur in liquid culture the myxamoebae are harvested normally in the exponential or stationary

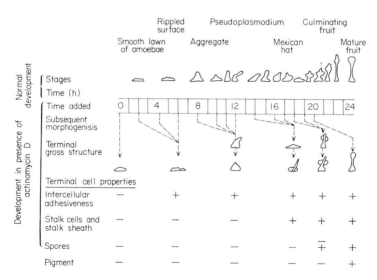

Fig. 2.5. The effect of actinomycin D on morphogenesis in *Dictyostelium discoideum*. From Sussman and Sussman (1969).

phases of growth, washed, resuspended and dispersed on non-nutrient solid surfaces to begin their morphological development. Several methods have been used though undoubtedly the method developed by Sussman and his co-workers has allowed a maximum control of synchrony. In this method the washed myxamoebae are plated on to 47 mm diameter Millipore filters or Whatman No. 50 filter paper in contact with buffered salts/Streptomycin solution. Under suitable conditions the developmental cycle can be completed in a highly synchronous manner within 24 h (Fig. 2.5). By means of this system it has been possible to examine the influence of antimetabolites, such as actinomycin D and cycloheximide, on the developmental sequence. Such compounds can be added at variable time intervals and may also be washed free. As a result much valuable information has been obtained on the timing of transcription and translation (Sussman and Sussman, 1969).

2.3 Cell Aggregation

The phenomenon of cell aggregation is of major importance in most eukaryotic organisms. In particular, it occurs in the initial stages of development of all animal embryos while failure to remain aggregated is a characteristic of certain cancer cells in animals. In the slime moulds one of the earliest manifestations of the initiation of differentiation is the aggregation of the myxamoebae marking the end of the interphase. The property of aggregation has been intensively studied with *Dictyostelium discoideum* and can be conveniently sub-divided into two aspects, chemotaxis and cell adhesion. Mutants that are defective in the property of aggregation are unable to complete the normal life-cycle.

Chemotaxis

It has been known for many years that the attraction which develops between myxamoebae is of a chemical nature. The aggregation of the myxamoebae does not require direct contact between cells since the attractant is operative up to a distance of 200μm whereas the diameter of a single myxamoeba rarely exceeds 15 μm. The insertion of non-porous materials such as glass, mica or cellophane between the centre of attraction and the myxamoebae prevents aggregation whereas a semi-permeable membrane allows free diffusion of the attracting molecules. Such observations suggest that the attractant is a low molecular weight compound. It was called acrasin by Bonner and can be destroyed by a phosphodiesterase enzyme, acrasinase, secreted by the myxamoebae. Why the organism should produce an enzyme to destroy such an important regulatory compound has baffled researchers for many years. However, it is now apparent that acrasinase is important in controlling the concentration of acrasin around the myxamoebae. Acrasin is responsible for orienting the myxamoebae which move towards areas of high concentration.

For many years research groups in many parts of the world vainly sought after the chemical identity of acrasin. Various steroids and alkaloids were implicated but none could be shown categorically to be as effective as the natural acrasin. During this period it has been shown that bacteria could attract myxamoebae and that cell-free supernatants from centrifuged bacteria were also active. All Gram-positive and Gram-negative bacteria so far tested can attract myxamoebae and the attractant of the bacterium *Escherichia coli* can activate several slime moulds including *Dictyostelium discoideum*. Myxamoebae were most sensitive to the attractant when they were close to their aggregation phase and high concentrations of the bacterial attractant could disperse aggregates even when they were in an advanced stage of morphogenesis. Subsequently, it has been shown that a low molecular weight compound cyclic 3′,5′-adenosine monophosphate (cyclic AMP) is extraordinarily active as an attractant to the myxamoebae and can be extracted from *E. coli*. Substantial amounts of cyclic AMP can be synthesized by *Dictyostelium* and there is now no doubt that cyclic AMP and acrasin are the same compound. Thus cyclic AMP appears to have at least a dual role in the life cycle of *Dictyostelium*, first as the attractant for hunting the bacterial prey, and secondly as a means of aggregating the free living myxamoebae.

Cyclic AMP is a compound found throughout the animal kingdom and is involved in the regulation of enzyme activity under the influence of hormones (Jost and Rickenberg, 1971). Many hormones stimulate the accumulation of cyclic AMP by accelerating the reaction

$$ATP \rightarrow cAMP + PPi$$

The enzyme catalysing this reaction, adenyl cyclase, is found in the plasma membrane of animal cells, while another enzyme, phosphodiesterase, found mainly in the soluble fraction of the cell can hydrolyse the cyclic AMP to 5′-adenosine monophosphate (Fig. 2.6). This is the only known physiological mechanism for degradation of cyclic AMP. Cyclic AMP has now been implicated in a two messenger hypothesis of hormone action in higher organisms. The hormone is the first messenger circulating in the blood, then binding to the plasma membrane of the target cell and activating adenyl cyclase. Cyclic AMP is

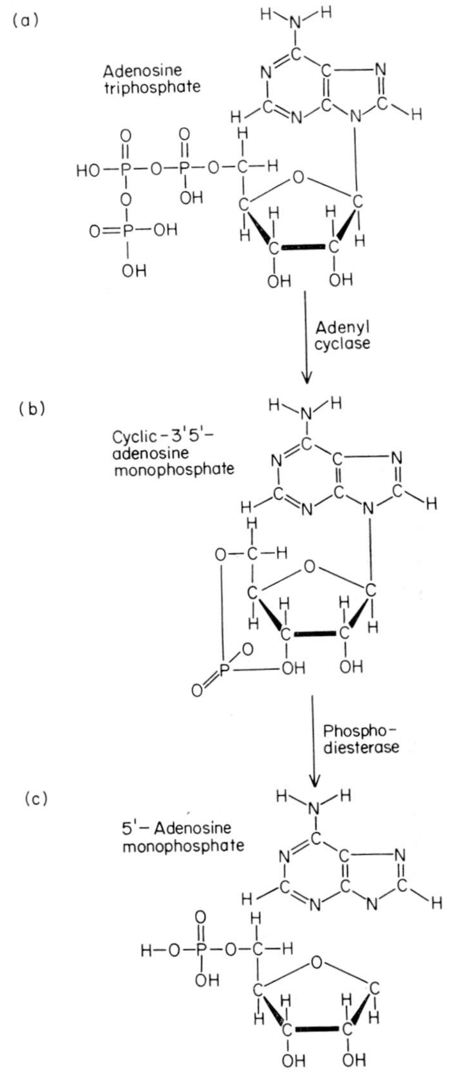

Fig. 2.6. The synthesis and breakdown of cyclic AMP. Cyclic AMP (b) is formed from the more familiar adenosine triphosphate, or ATP (a), in a one-step reaction catalysed by the enzyme adenyl cyclase. The word "cyclic" refers to the ring shape of the molecule's phosphate group. The substance is converted into 5′ AMP (c) by a phospho-diesterase that is specific for the reaction. From Bonner (1969).

considered to be a second messenger generated on the inner surface of the cell membrane, diffusing through the cell and activating the appropriate response. It is not clear what role this compound may play in eukaryotic differentiation since it does not seem to be able to activate quiescent genes but rather to modulate the activity of existing enzymes and to increase the rate of synthesis of pre-existing species of RNA and protein.

Cyclic AMP has also been found in *Euglena*, yeast and other amoebae. In bacteria it appears to be involved in controlling the synthesis of many inducible enzymes. It is not in itself an inducer, but acts only when the specific inducer is present. In the slime moulds cyclic AMP, over and above its internal role, acts as a hormone since it is extracellular and can provide communication over large intercellular spaces between myxamoebae.

When the myxamoebae of *Dictyostelium discoideum* begin to aggregate they move towards a centre point not by a steady movement but rather by a series of pulses. Cyclic AMP is now considered to be responsible for orienting the myxamoebae. Thus it is suggested that the pulsating movement of the myxamoebae is in response to the periodic release or secretion of cyclic AMP into the environment. The pulses or puffs of cyclic AMP leaving the initial myxamoebae (i.e. those myxamoebae which react first to nutrient exhaustion) elicit a pulse from neighbouring cells with approximately a 15 s lag and these move toward the original source. The amoeboidal movement lasts for about 100 s while the pulses are about 300 s apart. While a myxamoebae is moving, it is apparently refractory to further stimulation. This may arise by saturation of a responsive centre. Thus by means of a signal and response reaction there is a generation of an outward propagation of a wave of inward movement. In this way it is possible to explain how myxamoebae can move towards each other and then in streams moving towards the original signal source (Cohen and Robertson, 1971). This hypothesis has a sound theoretical basis but will require highly sophisticated experimentation to produce positive evidence. How does cyclic AMP elicit this effect? The answer to this question is not yet apparent but there may be close similarities between this mechanism and the chemical signalling in insects where it has been possible to show a relationship between emission rate of key molecules and the sensitivity threshold of the receiving organism. Enzymatic deactivation of the key molecules has been implied in the insect system while in *Dictyostelium* the enzyme phosphodiesterase appears to be mandatory in the controlling action of cyclic AMP.

If cyclic AMP is to function as an attractant then it is imperative that there must be a gradient of cyclic AMP concentration. The only way to achieve a gradient is by having the attractant consumed in the vicinity of the myxamoebae and so encouraging the myxamoebae to move outward to the higher concentrations of the gradient.

The myxamoebae of *Dictyostelium discoideum* secrete large quantities of a specific phosphodiesterase, and only when special care was taken to inhibit this enzyme or prevent it attacking the cyclic AMP was it possible to demonstrate the presence of cyclic AMP in *D. discoideum* (Fig. 2.6). This phosphodiesterase has now been identified as the specific acrasinase. Phosphodiesterase is produced throughout the vegetative, aggregative and migrative phases of the life cycle of *Dictyostelium discoideum*. Between the onset of aggregation and late aggregation there is a hundredfold increase in sensitivity to cyclic AMP. The sensitivity to phosphodiesterase also increases a hundredfold during aggregation. Thus during starvation a myxamoeba becomes highly sensitive to gradients of cyclic AMP and will move towards the source of secretion. If the source is another myxamoeba of the same strain aggregation will occur, while if the source is a bacterial colony feeding recommences.

A recent novel experiment involving the electrophoretic release of pulses of cyclic AMP from a microelectrode has induced and controlled normal aggregation of *Dictyostelium discoideum* (Robertson, Drage and Cohen, 1972). The precision of this technique has allowed a much more accurate time sequence of the onset of the chemotactic response during the interphase to be observed (Table 2.1). Cyclic AMP detection and movement response mechanisms are developed earlier in the interphase than the signal-replaying mechanisms. In addition the ability to

TABLE 2.1 Sequence of development of components of aggregation system of *Dictyostelium discoideum* during interphase. From Robertson, Drage and Cohen (1972).

Developmental event	Time after the onset of interphase	
	Present investigation	Earlier results
1. Weak chemotactic response to cyclic AMP	2 h	A few hours
2. Full chemotactic response to cyclic AMP	4 h	4h
3. Relaying of signal	6 h	Qualitative description
4. Appearance of autonomously signalling cells	6–8 h	6–8 h
5. Contact formation	8 h	9 h

form the intercellular contacts found in aggregating streams develops after the ability to propagate the cyclic AMP signal.

It has been demonstrated that exogenous cyclic AMP inhibits endogenous cyclic AMP synthesis thus implying that a feedback control of cyclic AMP production may be important during aggregation. Thus as soon as certain myxamoebae begin to secrete cyclic AMP they will prevent the neighbouring cells secreting the attractant. Such myxamoebae would become aggregation centres and would inhibit or destroy weaker centres by blocking their cyclic AMP production. Aggregation is also a function of calcium concentration and cyclic AMP induced aggregation is a calcium dependent phenomenon. Sudden increases in the concentration of extracellular cyclic AMP cause a large outflow of calcium ions from myxamoebae at the pre-aggregative and aggregative stages, and it has been considered that this may be part of the chemotactic action of cyclic AMP (Chi and Francis, 1971).

The observation that cyclic AMP directly affects stalk-cell differentiation (Bonner, 1970) could well suggest that this compound exerts a profound role in intracellular differentiation as well as in aggregation.

Cell Adhesion

The second important aspect of the aggregation process is the contact interaction of the myxamoebae. Pre-aggregating and aggregating myxamoebae undergo changes in cellular adhesiveness and cohesiveness (the term cohesiveness should be used to indicate cell–cell adhesions, and adhesiveness to indicate cell substratum adhesions). Feeding myxamoebae have very low cohesiveness, pre-aggregating myxamoebae are sufficiently cohesive to form clusters, while aggregating myxamoebae are strongly cohesive. Although chemotaxis is considered as the main directional force in promoting aggregation cellular cohesiveness may also be involved. Examination of myxamoebae in aggregating streams show that they are elongated in the direction of movement and their front and back surfaces remain in contact as they move. This phenomenon has been termed "contact following" and may also be important within the slug during later stages of development.

Clearly cohesion of the aggregating myxamoebae is essential for promoting morphogenetic interaction. The cohesion of the myxamoebae is specific and involves cell recognition and sorting out. It is considered that the clumping or cohesiveness of aggregating myxamoebae is due to a change of the cell surface; new surface antigens appearing at aggregation. Cohesiveness can be completely blocked with a univalent antibody prepared from antisera of aggregating myxamoebae. Although these myxamoebae fail to adhere, they can still respond to acrasin thus showing that chemotaxis and cohesion are totally separate phenomena. It has also been considered that the cohesion factors exist on the myxamoebae in the form of an equatorial ring (Gerisch, 1968). The adhesive material has yet to be identified although evidence suggests substances like mucopolysaccharides.

2.4   General Metabolism

Aggregation and subsequent morphogenesis are initiated in *Dictyostelium discoideum* by nutrient limitation. Since all nutrient uptake has stopped by interphase, endogenous materials must provide the only source of energy for the forthcoming phases of development. Protein and amino acids have been implicated as the main sources of energy during morphogenesis while glucose is used primarily for glycogen and cell wall synthesis. During development there is a high rate of breakdown of macromolecular reserve materials while gluconeogenesis is of little importance. The high rate of protein turnover allows for extensive protein utilization as an energy source, for maintaining the composition of critical proteins, and for the synthesis of new enzymes (Table 2.2).

TABLE 2.2. Changes in cellular components during differentiation. From Gustafson and Wright (1972)

Cell Component	% Original dry weight		
	Amoeba	Sorocarp	Used
Protein, amino acids	50	30	20
RNA	15	8	7
Lipid	15	8?	7?
Carbohydrate	9	9	0
DNA	3?	3	0
	92	58	34

Wright (1967) has considered that many differentiating systems, in particular *Dictyostelium discoideum* and the eggs of reptiles and birds are closed off from their environment and are self-sufficient. Such systems start with a fixed amount of endogenous carbon sources and undergo a "programmed" starvation, at the end of which morphogenesis has been accomplished. A prerequisite for organisms functioning in this way would be the ability to undergo a controlled breakdown of macro-molecular reserve materials. Compared with vegetative growth complex, morphological events must be more accurately programmed. The accumulation of structural material for the completion of specific events must occur in a precise and predictable sequence. This enforced endogenous metabolism is believed to give a much greater reproducibility to the order of the sequential events necessary for differentiation. Furthermore, the controlled depletion of reserve materials may actually trigger the various phases of development. A possible "fine-level" control of development by metabolic intermediates will be discussed later.

Extensive catabolism of endogenous reserves takes place after interphase and amoebae lose a considerable part of their original dry weight. The polysaccharide end-products of differentiation are derived mainly from glycogen. At aggregation 5% of the dry weight of the pseudoplasmodium is made up of soluble glycogen and approximately 1% by trehalose and glucose. At the end of differentiation there is little overall loss of carbohydrate material, but the pattern of distribution is different and it is possible to account for nearly all of the original carbohydrate in the following identifiable compounds: the disaccharide trehalose (glucose 1,1-α-D-glucoside) which accumulates in the spore and apparently serves as a carbon and energy source during germination; an acid mucopolysaccharide composed of galactose, N-acetyl-galactosamine and glucuronic acid associated exclusively with the spores in the sorocarp; soluable glycogen, and the cellulose-glycogen cell-wall complex which in mature sorocarps constitutes the integument of the stalk and is also present in the spore coats. The pattern of carbohydrate synthesis during differentiation is shown in Fig. 2.7.

Vegetative myxamoebae are facultative anaerobes while all developmental stages after aggregation are obligately aerobic. A complete glycolytic pathway has been demonstrated in *Dictyostelium discoideum* together with the existence of a pentose phosphate pathway which would give the $NADPH_2$ essential for the biosynthetic reactions associated with differentiation. The vegetative myxamoebae grow mainly at the expense of amino acids and proteins, and glucose neither increases the total cell yield nor decreases the doubling time. During growth of the myxamoebae glucose is converted to glycogen which is subsequently used as a source of carbon precursors for synthesis of the various polysaccharides which accumulate during sorocarp maturation (Baumann and Wright, 1968).

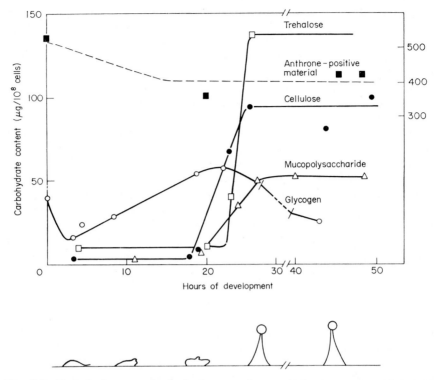

Fig. 2.7. Carbohydrate synthesis during morphogenesis in *Dictyostelium discoideum*. The experiments were performed with cells incubated on non-nutrient agar. From Sussman and Sussman (1969).

There is a high rate of protein degradation over the developmental period during which time 80% of the protein is newly synthesized, i.e. formed by turnover. At the same time there is turnover of RNA in the course of which 75% of the ribosomes present at the end of vegetative growth are degraded and replaced by new ribosomes (Table 2.2.). There is also a net loss of protein and of RNA during development which is presumably used for energy generation and also towards formation of polysaccharides.

2.5   Intracellular Control Mechanisms of Differentiation

Studies with Two-Membered Cultures

Since differentiation of myxamoebae can only occur after nutrient exhaustion it has been considered that growth and cell differentiation are mutually exclusive phenomena. Most biochemical studies on *Dictyostelium discoideum* have been concerned with the changes that occur in metabolism from aggregation onwards and contrasting biochemical concepts of differentiation have been vigorously propounded.

As with other organisms differential gene activity is invariably associated with differentiation in *Dictyostelium discoideum*. During differentiation the myxamoebae require novel proteins for specific cellular functions and this will entail the activation of the genes regulating these enzymes. The control of development at the cellular level has been the subject of much controversy, with one group advocating definite time steps for transcription and translation, while another group considers that control is exerted by substrate and product concentrations involving feedback inhibition or stimulation. Clearly, each model will contribute to a better understanding of the problem of differentiation, but surely the final analysis of differentiation cannot be adequately explained by either approach alone, but rather by a combination of both.

Sequential Enzyme Development

Sussman and his colleagues consider that the primary motivating force in differentiation is the control exerted by a series of ordered transcriptional and translational events which produce a new spectrum of enzyme activities. They have shown that many enzymes only appear during the developmental phase of *Dictyostelium discoideum* and such observations have led them to consider the concept of a developmental programme that controls the temporal and spatial synthesis or destruction of enzyme activity and hence morphogenesis (Sussman and Sussman, 1969).

In vitro enzyme determinations. If an enzyme is to be considered to be developmentally regulated and its synthesis controlled by a developmental programme then the following criteria should be met (Quance and Ashworth, 1972):

1. The specific activity should increase significantly during morphogenesis. 2. The increase in specific activity should be aberrant in morphologically deranged mutants. 3. The increase in specific activity should be influenced by known inhibitors of protein synthesis such as cycloheximide and actinomycin D. 4. This increase in specific activity is affected by mechanical disruption of the normal morphogenetic sequence. All enzymes that have been considered to be developmentally controlled obey 1–3 and where tested also conform to 4.

Morphological changes can be correlated with the appearance of a number of enzymes which are either not found, or are present in trace levels in the myxamoebae: N-acetylglucosaminidase, trehalose phosphate synthetase, threonine dehydrase, uridine diphosphoglucose pyrophosphorylase, uridine diphosphogalactose polysaccharide transferase, β-glucosidase, alkaline phosphatase, acid phosphatase and α-mannosidase (Sussman and Sussman, 1969; Loomis, 1969). These enzymes are correlated with development in so much that they change in measurable activity during development. Mutants of *Dictyostelium discoideum* with visibly abnormal development are also unusual in patterns of enzyme activity. Four of these enzymes will be considered in more detail.

N-acetylglucosaminidase. This enzyme catalyses the reaction:

p-nitrophenol N-acetylglucosamine $+ H_2O \longrightarrow p$-nitrophenol $+$
N-acetylglucosamine

The specific activity of this enzyme is very low in interphase myxamoebae, increases more than tenfold during the first 10 h of differentiation and then remains essentially constant throughout (Loomis, 1969).

Trehalose 6-phosphate synthetase. This enzyme is concerned with the formation of trehalose and cannot be detected in vegetative myxamoebae. Enzyme activity is first detected during aggregation, increases linearly up to 17 h of development and then decreases prior to stalk formation (Roth and Sussman, 1968).

Glucose 6-phosphate $+$ UDPG $\longrightarrow$ trehalose 6-phosphate $+$ UDP
trehalose 6-phosphate $+ H_2O \longrightarrow$ trehalose $+$ Pi

Uridine diphosphogalactose: polysaccharide transferase. This enzyme catalyses the transfer of galactose from uridine diphosphate galactose to a polysaccharide acceptor which is probably a precursor of a component of the slime sheath (Sussman and Osborn, 1964). The enzyme can be first detected during advanced pseudoplasmodium formation (12 h), it then accumulates for a further 8-h period and then is rapidly excreted into the external environment where the enzyme is destroyed. However, the product of this reaction, mucopolysaccharide, remains within the cells.

Uridine diphosphoglucose pyrophosphorylase. This is the enzyme responsible for catalysing the formation of uridine disphosphate glucose, a major metabolite in carbohydrate synthesizing reactions including trehalose formation (Ashworth and Sussman, 1967).

Glucose 1-phosphate $+$ UTP $\longrightarrow$ UTPG $+$ PPi

A low level of activity of this enzyme can be detected in vegetative myxamoebae and may represent a residual carry-over from the spore.

Enzyme activity increases linearly over an 8-h period after the end of cell aggregation. Maximum activity coincides with the terminal stages of development. The activity associated with the stalk cells disappears while that associated with the spore is retained.

A major criticism that must be raised with all developmental enzyme determinations has been concerned with the reliability and reproducibility of *in vitro* determinations. *In vivo*, the activity of a given enzyme concentration is dependent on substrate and co-factor availability, presence or absence of inhibitors and activators and the ionic environment. Furthermore, specific activity of an enzyme under conditions of optimum substrate concentration, as will occur in assays *in vitro*, only measures changes in the enzyme concentration. Again, in studies *in vitro* the physiological system is massively disrupted by the extraction procedures, and in particular damage to organelles such as nuclei, mitochondria and lysosomes may cause gross artefacts by allowing enzymes and metabolites to meet which are normally compartmentalized and kept apart. A more expansive discussion of the problems inherent in studying enzymes in crude extracts from differentiating cells is given by Sussman and Sussman (1969) and Gustafson and Wright (1972).

Studies on UDP-glucose pyrophosphorylase exemplify the problems that can occur with developmental enzymes. Ashworth and Sussman (1967) initially reported a tenfold increase during development followed by a decrease. Wright and Dahlberg (1968) considered that the increase in activity of this enzyme was an artefact of differential stability. They showed that the enzyme extracted from an early stage of development was much less stable than that extracted from a later stage and possessed a lower activity *in vitro*. However, Edmundson and Ashworth (1972) have recently devised a new assay procedure which categorically demonstrates that the original observations of Ashworth and Sussman (1967) for this enzyme were authentic.

REGULATION OF ENZYME ACCUMULATION

The increase in activity of all developmental enzymes shows sensitivity to the concomitant inhibition of protein synthesis and prior inhibition of RNA synthesis. These studies have centred around the use of cycloheximide which inhibits protein synthesis reversibly and with actinomycin D which specifically inhibits DNA-dependent RNA synthesis. Actinomycin D has become widely used for testing the relevance of transcription to a particular differentiation process (for a more detailed discussion on actinomycin D see Chapter 1). It is thought that the time at which actinomycin D treatment prevents the subsequent appearance

of the characteristic step in differentiation may in fact be the time
when the relevant information is being transcribed from the genome.
In these studies, rates of enzyme synthesis were measured and inhibitors
were added to the culture only after the proliferative stage had been
completed and the residual bacterial food supply removed by centrifugation.

The specific activity of several enzymes has been previously shown to
vary significantly during differentiation, and experiments with both
actinomycin D and cycloheximide suggest that the increase in specific
activity of at least seven enzymes (Fig. 2.8) requires concomitant

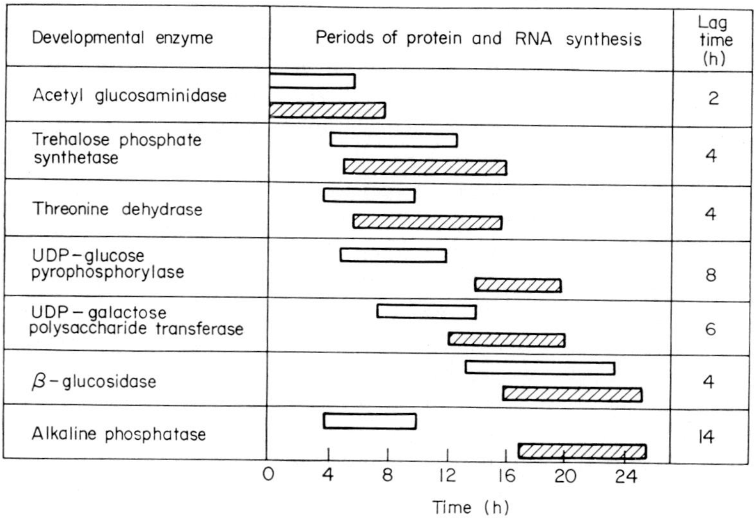

Developmental enzyme	Periods of protein and RNA synthesis	Lag time (h)
Acetyl glucosaminidase		2
Trehalose phosphate synthetase		4
Threonine dehydrase		4
UDP–glucose pyrophosphorylase		8
UDP–galactose polysaccharide transferase		6
β–glucosidase		4
Alkaline phosphatase		14

Time (h)

Fig. 2.8. Transcription and translation in development of *Dictyostelium discoideum.*
A schematic summary of the periods of RNA synthesis (open bars) and protein
synthesis (filled bars) required for accumulation of seven developmentally controlled
enzymes. Lag times were measured from the midpoint of the RNA synthetic period to
the midpoint of the protein synthetic period for each developmentally controlled
enzyme. From Loomis (1969).

protein synthesis and prior RNA synthesis. Significant differences
occurred between the time required for RNA synthesis and the period
of accumulation of most developmental enzymes. It has been considered that the changes in specific enzyme activity probably represent
alterations in specific gene activity. The period of transcription of each
enzyme was delineated by incorporating actinomycin D at progressively
later times during the developmental sequence, and the results imply

that the events of transcription and their duration are under developmental control. Gustafson and Wright (1972) have strongly questioned the significance of the inhibitor studies, and it is obvious that the results can be interpreted in many ways.

It has been considered that much of the essential transcription for differentiation in *Dictyostelium discoideum* may occur prior to aggregation or at the end of the growth phase of the life cycle. Pannbacker and Wright (1966) using 10 μg actinomycin D/ml demonstrated that when the drug was added before the early stages of aggregation, development was inhibited, but it had no effect when added to aggregated and later stages of development. Sussman and associates have applied much larger concentrations of actinomycin D at the later stages of aggregation and have obtained gross derangement of development together with drastic depression of RNA synthesis. They have confirmed the synthesis of new mRNA during morphogenesis (Sussman and Sussman, 1969), although it is still to be shown conclusively that this mRNA synthesis is intimately involved in development and not merely serving for the maintenance of a specific protein in the differentiating system. The concentrations of actinomycin D used by Sussman to cause derangement of morphogenesis are several times higher than those needed for inhibition of RNA synthesis in other systems. This may be due to the fact that actinomycin D binds specifically to dGMP residues and the G + C content of *D. discoideum* DNA is very low. It should also be noted that since development can only occur on a solid surface the concentration of actinomycin D reacting with the slug may never be accurately known.

Using actinomycin D and Miracil D (a bacteriostatic and carcinostatic thioxanthenane) Hirschberg, Ceccarini, Osnos and Carchman (1968) have suggested that the various mRNAs required for the normal sequence of morphogenesis in *Dictyostelium discoideum* may be formed during logarithmic growth (Table 2.3). Relatively stable and long-lived mRNAs have been suggested in other eukaryotic cells, and it has been considered that the orderly sequence of morphogenesis in *D. discoideum* may be reflected in the differential stability of the mRNAs.

Thus Sussman and co-workers consider that enzyme patterns and levels are the primary point of control during morphogenesis, and that alterations in their concentrations are presumably reflected in an altered rate of synthesis of some material essential to differentiation.

Changes in metabolite levels. Contrary to the above considerations Wright and her co-workers consider that changes in enzyme activity may be irrelevant to the control of differentiation. She considers that it is difficult if not impossible to conclude from *in vitro* enzyme studies

TABLE 2.3. Comparative activity of inhibitors against proliferation and aggregation of *D. discoideum*. From Hirschberg, Ceccerini, Osnos and Carchman (1968)

Compound	Concentration causing 50% inhibition of proliferation (M)	Minimum concentration preventing aggregation (M)
Actinomycin D	2×10^{-8}	2×10^{-5}
Miracil D	2×10^{-6}	2×10^{-5}
Miracil D derivatives*		
AN 207	4×10^{-6}	2×10^{-5}
AN 216	1×10^{-6}	1×10^{-5}
AN 304	2×10^{-5}	7×10^{-5}
AN 305	$>5 \times 10^{-5}$	$>1 \times 10^{-4}$
AN 316	$>2 \times 10^{-5}$	$>1 \times 10^{-4}$
AN 317	2×10^{-6}	1×10^{-5}
Quinacrine	$>1 \times 10^{-5}$	6×10^{-5}
Cycloheximide	4×10^{-5}	$>1 \times 10^{-4}$
Puromycin	Not done	$>1 \times 10^{-4}$

* Miracil　D = 1-diethylaminoethylamino-4-methyl-10-thioxanthen-9-one. The basic —$NHCH_2CH_2N$ $(C_2H_5)_2$ side chain of the parent compound is replaced by —$N(CH_3)CH_2CH_2N$ $(C_2H_5)_2$ in AN 207, by —$N(CH_3)CH_2CH_2N$ $(C_4H_9)_2$ in AN 216, and by —$N(CH_3)_2$ in AN 317; AN 305 is H_2—$NCH_2CH_2N(C_2H_5)_2$; AN 304 is 4-hydroxymethyl-Miracil D (Hycanthone); AN 316 is Miracil D sulfoxide.

alone (*a*) that an enzyme does not exist or occurs in another form or in a different place *in vivo*, or (*b*) what level of enzyme *in vivo* will actually limit the reaction (Wright, 1968). Furthermore, enzymes are normally present in vast excess *in vivo* compared to their substrates. Wright considers that changes in the concentration of key metabolites during differentiation are in themselves sufficient to cause the qualitative and quantitative changes in metabolite flux that occur during morphogenesis (Gustafson and Wright, 1972).

　　Wright (1966) has pointed out that the steady state concentration of key intermediates in the differentiating cells (such as ATP, UDPG, G-1-P) are well below the Km values for many of the enzymes for which they are the substrates. Thus changes in metabolic flux could be achieved by changes in substrate levels without altering the amount of the enzyme present. However compartmentalization may cause considerable localized changes in substrate concentrations and kinetic parameters of enzymes. The observation that there can be a considerable interval between transcription and product formation (Francis, 1969) would strengthen Wright's suggestion that many factors acting at all levels control product formation. Thus it becomes increasingly obvious that there is more to cell differentiation than the regulation of transcription and translation.

The extensive biochemical information now available on *Dictyostelium discoideum* has been used to devise a mathematical model of certain aspects of metabolism essential to differentiation (Gustafson and Wright, 1972). From this kinetic model it has been concluded that an increase in the specific activity of UDPG pyrophosphorylase is not required to account for the observed increased rate of synthesis of UDPG during development. The most important regulating factor in this system is the availability of glucose-1-phosphate and this further emphasizes the importance of measuring concentrations of precursors in differentiating systems. The weakness of this model is that it is based on a diagram of biochemical changes which represents a considerable oversimplification of the actual available data.

Gene-Product-Unit (GPU)

The contrasting viewpoint and experimental approaches of Sussman and Wright can be brought closer by considering differentiation to be controlled by the interactions that can exist between the gene, the product of gene action, and all the intermediate reactions associated with the end result. In an excellent review on the conflicting opinions of cellular control of differentiation, Francis (1969) considers that the state of a cell can be best described by considering the range of gene product units (GPU) which are active in the cell together with the interactions among these units. The GPU is the unit of metabolic action formed by a gene together with the mRNA, the enzyme formed, and the product of the enzymatic reaction associated with it (Fig. 2.9.). Such GPUs can interact in many ways and these include all the ways in which one part of the metabolizing system can affect another. Since differentiation consists of a succession of different metabolic states then each state may be described by their active GPUs and subsequently by their interactions cause one state to change to the next. If the activity of one GPU is obligatory for the subsequent change in activity of another GPU then it may be considered that the two GPUs are indeed part of the timing sequence which determines the order of events. As yet we do not know which GPUs make up the timing sequence. Furthermore, will the arrangement of GPUs in a timing sequence be a linear or complex branching form?

The studies of Sussman were primarily aimed at preventing transcription and translation activity in a GPU which could then break interactions between GPUs and thus reveal timing sequence links. On the

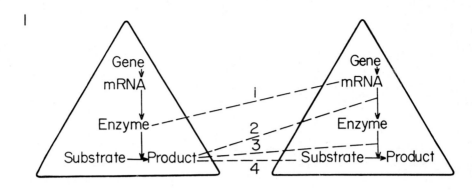

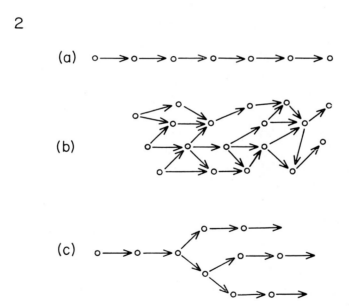

Fig. 2.9.1. Two GPU's and some possible modes of interaction between them. (1) Induction or depression; (2) "masking" of messenger RNA; (3) allosteric inhibition or competitive inhibition; (4) product of GPU (*a*) equals substrate of GPU (*b*).
2. Three examples of timing sequences. (*a*) Linear, (*b*) complex, and (*c*) branching timing sequences. Each hollow circle corresponds to a GPU, and each arrow represents the influence of one GPU on the action of another. From Francis (1969).

other hand Wright's concepts would consider that the important measure of the activity of a GPU is the rate at which the product is formed since it is the product which will determine the features of the cell which are of biological significance. It may not be important to cause gene transcription or changes in *in vitro* enzyme activity to be sequenced for differentiation since factors acting after these steps can change the order of appearance of products. The observation that there is a spread in time from transcription to product formation would favour Wright's suggestion that many factors acting at all levels control product formation.

In essence these studies do further indicate that differentiation does not depend solely on one essential controlling step but on many. The excellent review by Francis (1969) should be consulted to give a fuller understanding of GPUs and timing sequences in developmental programmes.

Only by understanding cellular timing sequences can we hope to achieve a major understanding of differentiation.

Studies of Axenic Culture

Sussman and Sussman (1967) first described a complex axenic medium containing liver extract and foetal calf serum which was capable of supporting growth and development of *Dictyostelium discoideum*. Watts and Ashworth (1970) have recently isolated a mutant strain of *D. discoideum* Ax-2 which will grow on a relatively simple axenic medium while yet retaining an ability to grow on bacteria.

Many chemical differences have been detected between myxamoebae of *Dictyostelium discoideum* NC-4 grown on bacteria and the mutant strain AX-2 grown axenically (Ashworth and Watts, 1970). DNA content of axenically grown myxamoebae is less than that of bacterially grown cells. However, it is considered that as much as half the DNA in the cells grown on bacteria is of bacterial rather than of myxamoebal origin. This observation must obviously complicate labelled nucleotide studies. The ability of the axenically grown myxamoebae to use glucose as an energy source must also question the role of proteins and amino acids in the growth of myxamoebae.

The specific activity of several enzymes considered to be part of a "developmental programme" have been measured during development after growth in axenic culture (Quance and Ashworth, 1972). The pattern of synthesis of these enzymes either remained the same whether grown axenically or bacterially (alkaline phosphatase, UDP glucose pyrophosphorylase and UDP galactose polysaccharide transferase), or was qualitatively similar but quantitatively affected (acid phosphatase, β-glucosidase) or both qualitatively and quantitatively affected by changes in the myxamoebae (α-mannosidase, N-acetylglucosaminidase).

Thus it has been possible with this new system to study the biochemical events that occur in two different cell populations going through the same morphogenetic sequences but having initially different chemical and enzymic compositions (Ashworth and Quance, 1972). Clearly, a given morphogenetic stage can contain a variety of enzymes depending on the origin of the myxamoebae from which it is derived. The view that cell differentiation is essentially a fixed process of selective gene activation is thus a partial and incomplete definition of the problem, and the "developmental programme" must involve events other than the sequence in which genes are activated and the amount of products that are produced from such genes (Quance and Ashworth, 1972).

Thus intracellular controls exist at the level of transcription and translation (coarse control) and also at the level of enzyme activity (fine control). The major problem posed by cell differentiation is the interaction between these two levels of control (Ashworth, 1971).

2.6 Intercellular Control Mechanisms of Differentiation

In multicellular eukaryotic organisms there is a distribution of functions in different parts of the developing organism. In *Dictyostelium discoideum* it has long been known that the number of prestalk cells and the number of prespore cells show a constant relation to each other and also that this relation can readjust itself if the aggregate is disrupted. It is also possible up to a certain phase in development for the prestalk and prespore cells to be interconvertible (Fig. 2.10). Thus there must exist some method of communication between the various parts of the developing cells since the posterior cells "know" how many anterior cells there are and vice versa. Many concepts have been tentatively put forward but as yet no single one can explain this remarkable phenomenon (for general discussion see Bonner, 1971; Ashworth, 1971).

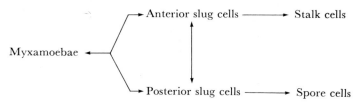

Fig. 2.10. Cell differentiation in *Dictyostelium discoideum*. From Ashworth (1971).

A control of enzyme synthesis by cellular interaction has been observed during development of *Dictyostelium discoideum* (Newell Longlands and Sussman, 1971). When partially developed aggregates were disaggregated to single cells and replated, they reaggregated, quickly recapitulated their previous morphological development and then completed sorocarp construction at almost the normal timing. However, during this time they also repeated the programme of enzyme accumulation forming an additional 300 units of UDP-glucose pyrophosphorylase activity regardless of the amount that had accumulated before disaggregation. The addition of actinomycin D at disaggregation inhibited further synthesis of the enzyme but did not inhibit the formation of the enzyme when added at a corresponding time to undisturbed developing cells. Disaggregated cells which were prevented from reforming normal cell contacts by EDTA formed little further enzyme. These experiments clearly show the need for cellular interaction to maintain the synthesis of this developmental enzyme even if the messenger RNA for the enzyme has already been formed. The reaggregation phenomenon is not a dedifferentiation and re-differentiation phenomenon but is in fact a morphological recapitulation of a previous phase of differentiation. This is in full agreement with Takeuchi (1969) who showed that ^{3}H thymidine-labelled prespore cells when mixed in the correct ratio with unlabelled prestalk cells sorted themselves out during reaggregation so that the unlabelled cells regained their anterior position typical of prestalk cells and the labelled cells the posterior region typical of prespore cells.

Summary

Dictyostelium discoideum differs from most other eukaryotic organisms in that it undergoes a period of vegetative growth—feeding and cell division—as separate independent cells. Development is initiated by the removal of nutrients and can be inhibited (up to culmination stage) by the addition of nutrients. The free-living myxamoebae aggregate forming a tissue-like assemblage of cells, the pseudoplasmodium, composed of many thousands of myxamoebae; one-third becoming vacuolated stalk cells and two-thirds becoming the spores of the final erect fruiting body or sorocarp. The pseudoplasmodium can move about and responds to many external stimuli. The outer cellulose sheath has been implicated in the control of communication between distal parts of the pseudoplasmodium. The cells of the anterior end of the pseudoplasmodium turn into stalk cells while the cells at the posterior end turn into spores. Unlike other aerial reproductive structures in the fungi there is no evidence for mitotic divisions at the apex of the stalk. The stalk is merely a vehicle for the spores to climb up out of their subterranean habitat.

During vegetative feeding the myxamoebae feed on bacteria and appear to be oriented and attracted towards them by an extracellular bacterial attractant, acrasin. Upon starvation the cells enter the new social phase and this appears again to be promoted by the secretion of acrasin by the myxamoebae. Acrasin can be destroyed enzymatically by acrasinase secreted by the myxamoebae. Acrasin has now been identified as cyclic AMP, a compound which now appears to have major involvement in cellular metabolism in most organisms. How myxamoebae respond to cyclic AMP is still not well understood but may resemble the mode of the action of chemical attractants of insects and chemotactic hormones such as sirenin (see Chapter 7). A better

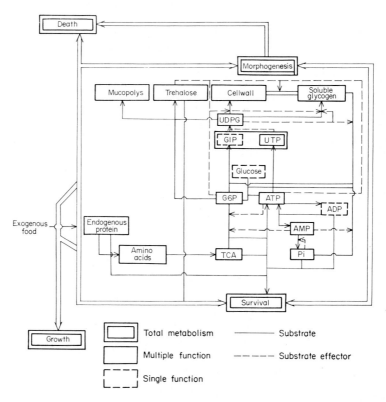

Fig. 2.11. Some of the interdependent and competing pathways necessary to both survival and morphogenesis in *Dictyostelium discoideum*. From Wright (1968).

understanding of one of these mechanisms may well lead to an understanding of them all. The biochemical steps in this process are undoubtedly extremely subtle and interwoven and will make ultimate understanding difficult. Clearly, an eventual solution can only come by biochemical and biophysical experimentation coupled with whole organism biology. Present evidence would suggest that the attractant is emitted in pulses and that gradient development is enhanced by the consumption of the attractant by acrasinase activity at the cell surface.

Differential gene activity is invariably associated with cellular differentiation in this organism. Major changes occur in macromolecular synthesis during development. How does the organism know when and how to control gene action? This problem has been vigorously discussed from several directions. Is development controlled by definite time steps for transcription and translation, or is it controlled by substrate and product concentrations involving feedback inhibition or stimulation? Clearly both concepts are correct and the answer lies in the interpretation. There can be no one key controlling step in development but rather an ordered sequence of control mechanisms. The identification of these mechanisms is undoubtedly one of the most challenging aspects of developmental biology.

Dictyostelium discoideum bridges the gap between the unicellular and the multicellular eukaryotic organisms and clearly shows the three main aspects of multicellular development, viz. differentiation, pattern formation and morphogenesis. Differentiation can be considered as the gene-activated intracellular processes that cause the appearance within the cell of certain biochemical or cytological characteristics. Pattern formation involves the spatial organisation of differentiation in so much that the different cell types achieve the correct spatial, temporal and proportional relationships to each other. Morphogenesis is concerned with the shape and form of the organism.

Recommended Literature

General Reviews

Ashworth, J. (1971). Cell development in the cellular slime mould *Dictyostelium discoideum*. *Symposium of Society of Experimental Biology* **25,** 27–49.

Bonner, J. T. (1967). "The Cellular Slime Moulds". Ed. 2. Princeton University Press, Princeton.

Bonner, J. T. (1969). Hormones in social amoebae and mammals. *Scientific American* **220,** 78–91.

Bonner, J. T. (1971). Aggregation and differentiation in the cellular slime moulds. *Annual Review of Microbiology* **25,** 75–92.

Francis, D. (1969). Time sequences for differentiation in cellular slime molds. *Quarterly Review of Biology* **44**, 277–290.

Garrod, D. and Ashworth, J. M. (1973). Development of the cellular slime mould *Dictyostelium discoideum. Symposium Society of General Microbiology* **23**, 407–435.

Gustafson, G. L. and Wright, B. E. (1972). Analysis of approaches used in studying differentiation of the cellular slime mold. *Critical Reviews in Microbiology* **1**, 453–478.

Sussman, M. and Sussman, R. R. (1969). Patterns of RNA synthesis and enzyme accumulation and disappearance during cellular slime mould cytodifferentiation. *Symposium Society General Microbiology.* **19**, 402–435.

Wright, B. E. (1966). Multiple causes and controls in differentiation. *Science* **153**, 830–837.

Wright, B. E. (1967). On the evolution of differentiation. *Archiv für Mikrobiologie* **59**, 335–344.

Cultural Techniques

Sussman, R. R. and Sussman, M. (1967). Cultivation of *D. discoideum* in axenic medium. *Biochemical and Biophysical Research Communications* **29**, 53–57.

Watts, D. J. and Ashworth, J. M. (1970). Growth of myxamoebae of the cellular slime mould *Dictyostelium discoideum* in axenic culture. *Biochemical Journal* **119**, 171–174.

Ultrastructure

Ashworth, J. M., Duncan, D. and Rowe, A. J. (1969). Changes in fine structure during cell differentiation of the cellular slime mould *Dictyostelium discoideum. Experimental Cell Research* **58**, 73–78.

George, R. P., Hohl, H. R. and Raper, D. B. (1972). Ultrastructural development of stalk-producing cells in *Dictyostelium discoideum*, a cellular slime mould. *Journal of General Microbiology* **70**, 477–489.

Genetics

Sinha, U. and Ashworth, J. M. (1969). Evidence for the existence of elements of a parasexual cycle in the cellular slime mould *Dictyostelium discoideum. Proceedings of the Royal Society, Series B* **173**, 531–540.

Chemotaxis

Bonner, J. T., Barkley, D. S., Hall, E. M., Konijn, T. M., Mason, I. W., O'Keefe, III, G, and Wolfe, B. B. (1969). Acrasin, acrasinase and the sensitivity to acrasin in *Dictyostelium discoideum. Developmental Biology* **20**, 72–87.

Chang, Y. (1968). Cyclic 3', 5'-adenosine monophosphate phosphodiesterase produced by the slime mold *Dictyostelium discoideum. Science* **160**, 57–59.

Chi, Y. and Francis, D. (1971). Cyclic AMP and calcium exchange in a cellular slime mold. *Journal of Cellular Physiology* **77**, 169–174.

Cohen, M. H. and Robertson, A. (1971). Wave propagation in the early stages of aggregation of cellular slime molds. *Journal of Theoretical Biology* **31**, 101–118.

Cohen, M. H. and Robertson, A. (1971). Chemotaxis and the early stages of aggregation in cellular slime moulds. *Journal of Theoretical Biology* **31**, 119–130.

Konijn, T. M., van de Meene, J. C. G., Chang, Y. Y., Barkley, D. S. and Bonner, J. T. (1969). Identification of adenosine-3', 5'- monophosphate as the bacterial attractant for myxamoebae of *Dictyostelium discoideum. Journal of Bacteriology* **99**, 510–512.

Konijn, T. M., Barkley, D. S., Chang, Y. Y. and Bonner, J. T. (1969). Cyclic AMP: A naturally occurring acrasin in the cellular slime molds. *The American Naturalist* **102**, 225–233.

Mason, J. W., Rasmussen, H. and Dibella, F. (1971). 3'5' AMP and Ca^{2+} in slime mold aggregation. *Experimental Cell Research* **67**, 156–160.

Robertson, A., Drage, D. J. and Dohen, M. H. (1972). Control of aggregation in *Dictyostelium discoideum* by an external periodic pulse of cyclic adenosine monophosphate. *Science* **175**, 333–335.

Cell Adhesion

Gerisch, G. (1968). Cell aggregation and differentiation in *Dictyostelium*. *Current Topics in Developmental Biology* **3**, 157–197.

General Metabolism

Ceccarini, C. and Filosa, M. (1965). Carbohydrate content during development of the slime mold *Dictyostelium discoideum*. *Journal of Cellular and Comparative Physiology* **66**, 135–142.

Wright, B. E. and Dahlberg, D. (1967). Cell wall synthesis in *Dictyostelium discoideum*. II. Synthesis of soluble glycogen by a cytoplasmic enzyme. *Biochemistry* **6**, 2074–2079.

Intracellular Control Mechanisms

Ashworth, J. M. and Quance, J. (1972). Enzyme synthesis in myxamoebae of the cellular slime mould *Dictyostelium discoideum* during growth in axenic culture. *Biochemical Journal* **126**, 601–608.

Ashworth, J. and Sussman, M. (1967). The appearance and disappearance of uridine disphosphate glucose pyrophosphorylase activity during differentiation of the cellular slime mold *Dictyostelium discoideum*. *Journal of Biological Chemistry* **242**, 1696–1700.

Ashworth, J. M. and Watts, D. J. (1970). Metabolism of the cellular slime mould *Dictyostelium discoideum* grown in axenic culture. *Biochemical Journal* **119**, 175–182.

Edmundson, T. D. and Ashworth, J. M. (1972). 6-Phosphogluconate dehydrogenase and the assay of uridine diphosphate glucose pyrophosphorylase in the cellular slime mould *Dictyostelium discoideum*. *Biochemical Journal* **126**, 593–600.

Hirschberg, E., Ceccarini, C., Osnos, M. and Carchman, R. (1968). Effects of inhibitors of nucleic acid and protein synthesis on growth and aggregation of the cellular slime mold *Dictyostelium discoideum*. *Proceedings of the National Academy of Sciences* **61**, 316–323.

Jost, J. and Rickenberg, H. U. (1971). Cyclic AMP. *Annual Review of Biochemistry* **40**, 741–774.

Loomis, W. F. (1969). Acetylglucosaminidase, an early enzyme in the development of *Dictyostelium discoideum*. *Journal of Bacteriology* **97**, 1149–1154.

Loomis, W. F. Jr. (1969). Developmental regulation of alkaline phosphatase. *Journal of Bacteriology* **100**, 417–422.

Pannbacker, R. G. and Wright, B. E. (1966). The effect of actinomycin D on development in the cellular slime mold. *Biochemical and Biophysical Research Communications* **24**, 334–338.

Quance, J. and Ashworth, J. M. (1972). Enzyme synthesis in the cellular slime mould *Dictyostelium discoideum* during the differentiation of myxamoebae grown axenically. *Biochemical Journal* **126**, 609–615.

Roth, R. and Sussman, M. (1968). Trehalose 6-phosphate synthetase (uridine diphosphoglucose: D-glucose 6 phosphate 1-glucosyltransferase) and its regulation during slime mould development. *Journal of Biological Chemistry* **243**, 5081–5087.

Sussman, M. and Osborn, M. (1964). UDP-galactose polysaccharide transferase in the cellular slime mold *Dictyostelium discoideum*: appearance and disappearance of activity during cell differentiation. *Proceedings of the National Academy of Sciences U.S.* **52**, 81–87.

Wright, B. E. (1968). An analysis of metabolism underlying differentiation in *Dictyostelium discoideum*. *Journal of Cellular Physiology* **72**, 141–160.

Wright, B. E. and Dahlberg, D. (1968). Stability *in vitro* of uridine diphosphoglucose pyrophosphorylase in *Dictyostelium discoideum*. *Journal of Bacteriology* **95**, 983–985.

Intercellular Control Mechanisms

Newell, P.C., Longlands, M. and Sussman, M. (1971). Control of enzyme synthesis by cellular interaction during development of the cellular slime mold *Dictyostelium discoideum*. *Journal of Molecular Biology* **58**, 541–554.

Takeuchi, I. (1969). *In* "Nucleic Acid Metabolism Cell Differentiation and Cancer Growth". Ed. Cowdoy, E. V. and Sena, S. Pergamon Press, Oxford.

3 A Plasmodial Slime Mould: *Physarum polycephalum*

3.1 Introduction

The plasmodial slime moulds are an unusual group of micro-organisms in so far that they display certain characteristics of growth typical of some animal and plant forms. The creeping, vegetative phase found in many species has undoubted parallels in the animal world while the phase of sexual reproduction, in having spores with cellulose walls, is plant-like. The slime moulds are now considered taxonomically to have fungal origins but to have departed sufficiently from the main evolutionary line of the true fungi to represent a separate subdivision—the Myxomycotina (Alexopoulos, 1962). In nature, the slime moulds which can be found growing mainly in soil in North Temperate regions exist by engulfing bacteria, other small forms of life, or organic debris. Over 400 species have now been classified.

Although no major ecological or economic role can yet be ascribed to these organisms, several industrial laboratories are exploring their potential as possible producers of enzymes and protein biomass. Because in some cases the large acellular vegetative thallus or plasmodium (up to 15 cm in diameter) may be considered biochemically as a single cell, they have been used extensively by cytologists, physiologists and biochemists to study many fundamental aspects of cellular metabolism. Furthermore, most slime moulds exhibit a remarkably precise mitotic synchrony during plasmodial growth. Because of this and the distinct stages of the life cycle they have become useful organisms for studying fundamental aspects of eukaryotic microbial growth and differentiation. There are several excellent recent reviews on various aspects of the biology of the slime moulds which should be considered for a fuller appreciation of these little known, but potentially exciting organisms. By far the most extensive and rewarding studies on the plasmodial slime moulds have been carried out with *Physarum polycephalum*, and the remainder of this chapter will be concerned almost entirely with this organism.

The life-cycle of *Physarum polycephalum* consists of alternating haploid and diploid generations. The haploid phase is limited to the uni-nucleate spore (10 μ in diameter) which germinates in liquid culture to form a flagellated swarm cell or in surface culture to give a non-flagellated amoeba (Fig. 3.1). In unfavourable conditions these amoebae

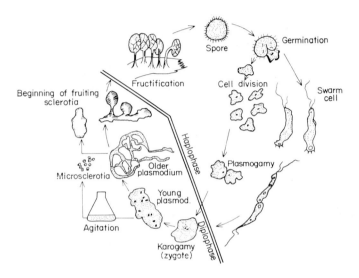

Fig. 3.1. Life-cycle of *Physarum polycephalum*. From Alexopoulos (1962).

or swarm cells can encyst and become resistant, re-emerging when the environmental conditions become more amenable for growth. In the presence of adequate nutrients the haploid cells will divide repeatedly and form a large population of similar cells. There have been few studies on haploid differentiation e.g. swarm amoeba cell or microcyst amoeba.

The plasmodial phase is initiated by a sexual process, the fusion of two amoebae. The two amoebae must differ in mating type, and the mating types, of which four have been identified, are determined by a series of alleles at a single locus.

A plasmodium may increase in size by the growth of a single zygote or it may enlarge by successive coalesces or fusions with other zygotes or developing plasmodia. Development in rapid shaken suspension is in the form of small microplasmodia containing between 2–200 diploid

nuclei, whereas development on a solid surface produces macroplasmodia containing up to several million diploid nuclei. The plasmodium
is a motile mass of protoplasm of variable form, without a rigid cell-wall,
and is not divided into cells, i.e. it is coenocytic. Within the plasmodium
the protoplasm streams to and fro and in many ways resembles a giant
multinucleate amoeba. A vigorous protoplasmic streaming takes place
continuously through a fine network of veins or channels at a rate as
fast as 1 mm/s with reversal of flow approximately each minute. In the
presence of adequate nutrients the plasmodium, although exhibiting
protoplasmic streaming, will remain more or less stationary. When the
nutrients start to become exhausted, however, the plasmodium starts to
move in pulsating waves over the surface of the medium and becomes
reticulate in appearance. The wandering ceases when a new food source
is encountered. It is not yet known what mechanisms initiate and control
the movement of the plasmodium.

The plasmodia of approximately 30–40 species of slime mould can be
grown in the laboratory, and in most cases this has only been achieved
with media containing dead or living bacteria. However, several species
can now be grown on semi-defined or completely defined chemical
media at the plasmodial stage although there is difficulty in getting
most slime moulds to complete their life-cycle in chemically defined
axenic culture. The extensive use of *Physarum polycephalum* as an experimental organism has resulted largely from its ease of axenic culture.
Production of large quantities of plasmodia normally occurs in shaken
flasks or large-scale fermenters.

Under unfavourable growing conditions, e.g. desiccation, low temperature, or starvation, the plasmodium of *Physarum polycephalum* will
cease to pulsate and may become encysted to form thick-walled resistant
structures; microplasmodia will form spherules and macroplasmodia
will form sclerotia. When sclerotia or spherules are returned to a normal
growth environment a reversion to a pulsating plasmodium takes place.
During sclerotia formation cell cleavage occurs without mitosis so wall
synthesis may be studied in the absence of nuclear division.

The sexual cycle of the slime mould is completed when the plasmodium differentiates into sporangia in which the haploid uninucleate
spores are formed. Sporangial formation marks the end of the life-cycle
and is the only stage where a cellular growth form occurs. Sporulation
requires exacting environmental conditions involving a period of
starvation in darkness in the presence of niacin followed in most species
by a brief exposure to visible light.

Thus there are three distinct and separate stages in the life-cycle of
Physarum polycephalum: a cyclic vegetative mitotic phase, a transient

vegetative phase of sclerotization or spherulation, and an irreversible sexual phase. Each phase offers a model system for experimental studies concerning specific features of growth and differentiation. These three forms of differentiation will be examined, in detail, using results obtained from *Physarum polycephalum*.

3.2   The Plasmodial Growth Phase

Until recently the culturing of slime moulds required the use of complex, undefined growth media. Such methods imposed obvious restrictions and disadvantages for obtaining meaningful results on comparative biochemical studies of morphogenesis; a situation not unlike that occurring with *Dictyostelium discoideum* (see Chapter 2). However, in recent years *Physarum polycephalum* and several other slime moulds have been successfully cultured axenically on defined or semi-defined media.

The microplasmodia of *Physarum polycephalum* are routinely maintained in the dark in semi-defined medium at 21–22°C in shaken flask culture. Most biochemical studies involve the use of surface cultures that are prepared by allowing microplasmodia to coalesce on moist filter paper or millipore membrane surfaces. When such cultures are incubated at 26°C the mitotic cycle of each developing plasmodium is approximately 8–12 h. A plasmodium with an initial diameter of about 1 cm will attain an approximate size of 8 cm in 24 h by which time it has completed three mitotic divisions. A novel technique in which giant plasmodia (up to 15 cm in diameter) can be grown has recently been developed (Mohberg and Rusch, 1969).

Vegetative growth in most eukaryotic organisms normally involves the mitotic division of the nucleus with concurrent formation of new cells. The mitotic cycle consists of a sequence of stages: a mitosis (M) to initiate the cycle, a phase of growth (G1), a phase of DNA synthesis (S), a second wave of growth (G2) and a mitosis (M) to terminate the cycle. In *Physarum polycephalum* the mitotic cycle does not follow this sequence lacking a demonstrable G1 phase with the cycle proceeding directly from M to S phase. In normal mitotic division the G1 period is considered to be the time in the cycle when specialized substances required for specific functions other than growth are synthesized; this period is found only in those cells that are somewhat specialized in some degree. The absence of a G1 phase is not uncommon in certain rapidly dividing eukaryotic cells, e.g. embryonic cells of frogs and sea urchins. When plasmodia are cultured with a mitotic cycle of 12 h, the S phase has a duration of about 3 h and the cell spends the remainder of the cycle in G2 (Fig. 3.2). The nuclei of *Physarum polycephalum* are small and

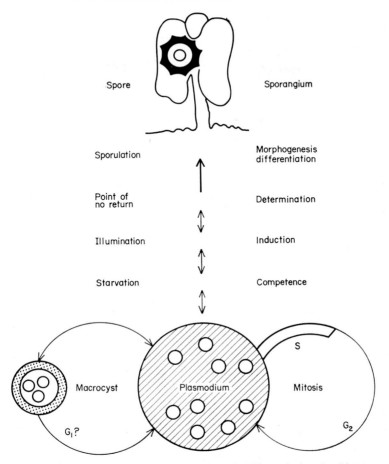

Fig. 3.2. Schematic representation of plasmodial differentiation in *Physarum poly-cephalum.* From Sauer (1973).

contain between 20 and 50 chromosomes. An unusual feature of mitosis in this organism and in all other fungi is that the nuclear membrane does not disappear during division (Fig. 3.3). In all other eukaryotic cells the nuclear membrane disappears early in mitosis and reforms again during telophase.

However, the outstanding importance of *Physarum polycephalum* to comparative biochemical studies on cellular development is that the plasmodium is coenocytic and all nuclear divisions occur with a natural synchrony. Furthermore, plasmodia are sufficiently large to allow detailed biochemical analysis to be made on individual synchronous

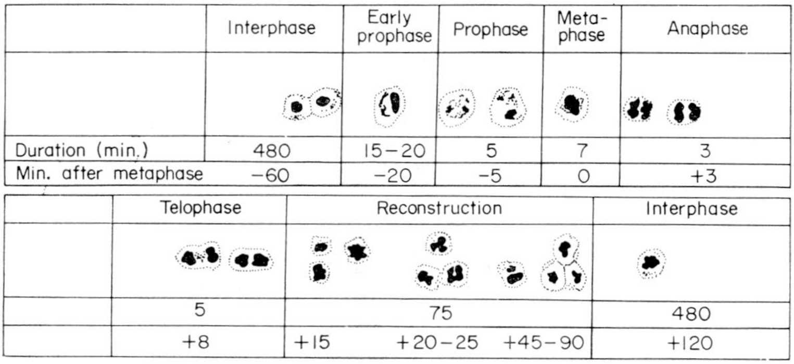

	Interphase	Early prophase	Prophase	Meta-phase	Anaphase
Duration (min.)	480	15–20	5	7	3
Min. after metaphase	−60	−20	−5	0	+3

	Telophase	Reconstruction			Interphase
	5	75			480
	+8	+15	+20–25	+45–90	+120

Fig. 3.3. Mitotic stages in the nucleus of the *Physarum* plasmodium. Alcohol-fixed plasmodial smears were photographed through a phase-contrast microscope. Representative nuclei on the photographs were traced to give the drawings of the figure. All drawings are at the same magnification. Stippled outlines of nuclei represent surrounding cytoplasm and not nuclear membranes. On the first line below each row of nuclei is the duration in minutes of each stage of mitosis, and on the second line is the number of minutes before the beginning or after the ending of metaphase. From Mohberg and Rusch (1969).

plasmodia. In other multinucleate eukaryotes total synchrony does not occur and although partial synchrony can be achieved in certain populations of unicellular organisms this can only occur by imposing considerable artificial environmental pressures. Thus, as a consequence of the natural synchrony in a large coenocytic cell it has been possible to make major advances in the understanding of the cellular processes of DNA replication, transcription and mitosis. The vast literature on this subject has been lucidly reviewed by Cummins (1969; Rusch, 1969 and Sauer, 1973).

The synthesis of DNA takes place immediately following mitosis and extends over a period of 3 h. Inhibition of DNA synthesis by fluorodesoxyuridine also inhibits mitosis. DNA molecules that are replicated during a given temporal segment in one S phase are replicated during a similar temporal segment of the following S phase. This was deduced from the observation that nuclei from S period plasmodia transplanted into G2 plasmodia continued DNA synthesis, but nuclei from plasmodia in the G2 period did not synthesize DNA when placed in S period plasmodia.

The G2 period apparently serves for the transcription and synthesis of substances essential for mitosis and for the enlargement of the plasmodium. There are now several lines of research which strongly imply that synthesis of factors which control the synchronization of the nuclear divisions occurs in the cytoplasm. Indirect evidence indicates that mitotic stimulation is brought about by the production of specific cytoplasmic diffusable protein factors late in G2. A schematic representation of the complete fusion experiment which has led to these conclusions is shown in Fig. 3.4. Each experiment utilized replicate discs cut from two plasmodia at different stages in the mitotic cycle. The discs cut from the plasmodium late in the cycle, i.e. closer to mitosis, were designated A and those from the plasmodium earlier in the cycle were designated B. Replicate discs from both A and B were fused by overlayering A with B to form the experimental fusion culture A/B as well as self-fused to form the fused controls A/A and B/B. The time of

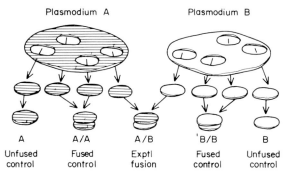

Fig. 3.4. A schematic representation of a fusion experiment depicting: the excision of replicate discs from plasmodia A and B: the maintenance of discs from A and B as unfused controls; the overlayering of discs from the same plasmodium to form fusion controls, A/A and B/B; and the overlayering of discs from different plasmodia to form the experimental fusion culture, A/B. From Chin, Friedrich and Bernstein (1972).

mitosis subsequent to fusion was determined in the experimental fusion culture A/B as well as the fusion control A/A and B/B and in unfused controls A and B.

The experimental results showed that: (i) all nuclei in all plasmodial discs in the series, fused or unfused, divided synchronously; (ii) synchronous mitosis in the experimental fusion culture A/B usually occurred at a time intermediate between the unfused control A and B and between the fused controls A/A and B/B. Thus nuclei from plasmodia early in the cell cycle were stimulated to divide sooner than they would otherwise have done, and nuclei from plasmodia late in the cycle were retarded. It has not yet been possible to isolate and fully characterize the factors which regulate mitosis but their importance cannot be overstated in molecular biological terms.

The levels of messenger RNA, transfer RNA and ribosomal RNA vary during the growth cycle. The last mRNA for mitosis is completed about 0·5 h before metaphase, whereas the last essential protein for this process is synthesized only 15 min. before metaphase. Finally, mitosis does not occur until after DNA has been replicated.

Thus during the growth phase of the plasmodium nuclear division and DNA synthesis provide structural and molecular markers for the cyclic acts of differentiation.

3.3   Encystment

In the normal developmental pattern of *Physarum polycephalum* the plasmodium will eventually cease growing, form a fructification and produce haploid spores. However, under a wide variety of environmental conditions the plasmodium can become encysted, being converted into an irregular hardened mass, the *sclerotium*, which is capable of prolonged dormancy yet retains the ability to rapidly regenerate and resume vegetative growth when suitable environmental conditions are resumed. Sclerotium formation is not uncommon in many true fungi and will be discussed again in Chapter 5.

More recent studies on the general phenomenon of sclerotization have examined a special type of sclerotization called spherulation in which microplasmodia growing in liquid shaken culture are induced to form *microcysts* or *spherules* by osmotic shock during active growth or by starvation. Thus under certain cultural conditions that can interfere with mitotic growth macroplasmodia will encyst to form sclerotia while microplasmodia will be transformed into spherules.

Sclerotization of actively growing and pulsating plasmodia can be induced in many ways (Jump, 1954):

Sclerotization by gradual desiccation: A portion of plasmodium is allowed to spread on a cellophane membrane in contact with a 2% agar gel surface in a Petri-dish. The cellophane membrane with the growing plasmodium is then stripped from the agar and placed in a dry Petri-dish at 25°C.

Sclerotization by starvation: As above but removed from nutrient agar to plain agar.

Sclerotization in hypertonic solutions: The supporting vehicle is dipped into 0·6 M sucrose for 5 min. and then transferred to an agar medium containing 0·6 M sucrose.

Sclerotization by exposure to sub-lethal concentrations of heavy metals or lower pH values:

Spherulation can easily be induced by either starvation or osmotic shock:

Spherulation by starvation: Actively growing microplasmodia are transferred to a non-nutrient mineral solution, and spherule formation takes place with good synchrony within 24–36 h. In this method there is a greatly increased slime production and this causes the clumping together of groups of spherules (Daniel and Baldwin, 1964).

Spherulation by osmotic shock: The addition of 0·5 M mannitol to a defined growth medium containing actively growing microplasmodia results in the formation of spherules within 35–65 h (Chet and Rusch, 1969). There is practically no slime production by this method and the spherules appear as single units and not as clusters. Mannitol added to a salts medium does not induce spherule formation. The morphological changes induced by starvation and osmotic shock are to all appearances similar.

In all forms of encystment there is a cessation of cytoplasmic streaming and a gelation of the plasmodium. Electron microscopic studies have shown that prior to sclerotization the plasmodium is divided into numerous subunits or spherules by line-up, enlargement and fusion of vesicles within the cytoplasm. An interspherule membrane is formed along the spherule membrane by stratification of fibrous material that appears to form in the vesicles and vacuoles. Associated with these processes there is a shrinkage of the nuclei to one-half diameter and an appearance of a Golgi apparatus or dictyosome. In the true fungi dictyosomes have only been observed in the Oömycetes.

The total amounts of RNA, protein and DNA decreased drastically during spherulation (Fig. 3.5), while there was a greater decrease in glycogen content during spherule induction by starvation than by polyol additions. Since slime production is considered to derive from the breakdown of glycogen this may account for the lower glycogen drop during polyol induction. The heavy slime production that occurred when polyols were added to mineral salts medium containing microplasmodia (no spherule production) would imply that slime production is the result of starvation and not the induction of spherule formation. The ability of low concentrations of cycloheximide to completely inhibit spherulation indicates an essential requirement for protein synthesis. Slime production does not appear to require new protein synthesis since it occurs in the presence of cycloheximide.

The ability to use different induction systems to achieve a similar morphological event has allowed exploration of the question of whether the activity pattern of certain enzyme markers would also be similar in both systems. Changing enzyme patterns have been documented during spherulation. In particular, glutamic acid dehydrogenase and a phosphodiesterase have been shown by Hüttermann and co-workers to

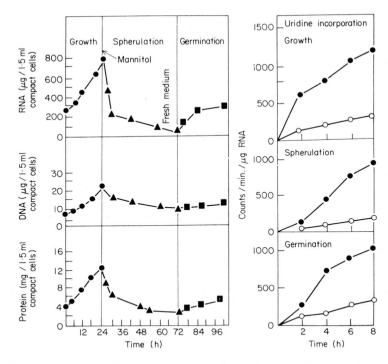

Fig. 3.5. Synthesis of RNA, DNA, and protein during growth, spherulation and germination. The cultures of *Physarum polycephalum* were transferred to synthetic medium containing mannitol after 24 h of growth. Mannitol induces spherulation. The spherules were again transferred to fresh medium 46 h after the beginning of spherulation. The graphs on the righthand side show incorporation of ³H-uridine into RNA during growth, spherulation and germination (●) and the effect of Actinomycin D on this incorporation (○). Chet and Rusch (1969).

increase up to 10-fold during spherulation induced by starvation. They have also shown that *de novo* synthesis of glutamic acid dehydrogenase does occur during this form of spherulation and that there is a change in isoenzyme patterns of phosphodiesterase. These studies have involved isopycnic centrifugation of deuterium-labelled extracts and conclusively demonstrate the *de novo* enzyme synthesis and imply that the enzyme changes were not promoted by allosteric control of small molecules or activation of molecules already present.

Both enzymes showed a much smaller increase in activity when spherulation was induced by mannitol. When the enzymes were assayed from microplasmodia in mineral salts plus mannitol medium (no induction of spherulation) there was little change in the activity of glutamic acid dehydrogenase but a large increase in the activity of phosphodiesterase. This would imply that the activity increase of phosphodiesterase mainly parallels starvation rather than induction and development of spherulation. The necessity of glutamic acid dehydrogenase for spherule formation is still unclear. The lack of a major increase in enzyme activity during spherule formation by osmotic shock may be because the materials which are provided by this enzyme and needed during spherulation are already present in the synthetic medium and consequently the increased synthesis of this enzyme is not triggered.

The role of transcription during spherule formation is still uncertain. Actinomycin D has little effect on the process although spherules formed in the presence of this nucleic acid antagonist have a low viability. Other studies have indicated that some species of RNA decrease while new species appear during spherulation. The significance of these observations is not yet understood. The possibility of a G1 phase occurring during spherulation has recently been considered since it has been demonstrated that the amount of DNA in spherule nuclei is approximately half the amount found in the nuclei of vegetative plasmodia. This could well be correct if it is considered that spherulation is an act of differentiation.

3.4 Sporulation

Sporulation has a threefold advantage to the organism: as a mechanism of survival, as a means of dispersal, and for genetic recombination during meiosis. During sporulation the entire plasmodium is converted into one or several fruiting structures (Fig. 3.1) and consequently the somatic and reproductive phases rarely occur simultaneously in the same plasmodium.

In common with most types of microbial sporulation depletion of nutrients is an obligate feature in initiating the events which lead to sporulation in *Physarum polycephalum*. In laboratory practice actively growing plasmodia are plated on to a nutrient-free surface (filter paper or Millipore filter) moistened with salts solution containing niacin and allowed to starve for four days. During this period the plasmodia move over the surface in search of food. There is a considerable decrease in the size of the plasmodium during this period as the processes of life exist by utilizing endogenous food reserves. Although there is at least one mitosis and concomitant nucleic acid synthesis during the starvation period there is a net decrease in DNA, RNA and protein. In the absence of niacin the subsequent stages of sporulation will not occur.

The exact role of niacin is not known but it may be related to the formation of pyridine nucleotides which may be involved in the subsequent photosensitive reactions. Light is necessary for induction of sporulation in pigmented plasmodia such as *Physarum polycephalum*, but not in non-pigmented plasmodia, *P. compressum*. The photosensitive state can only be induced by the prior 4-day period of starvation in the presence of niacin. Plasmodia that have been subjected only to starvation or to exposure to light will start to grow again when given adequate nutrients, and returned to the dark. The process of sporulation can be reversed up to a critical period of development by the addition of nutrients. However, after this critical period has been passed the plasmodium is irreversibly committed to sporulate even if returned to a suitable growth medium (Fig. 3.2).

There is an inverse relationship between the length of the period of illumination to initiate sporulation and age of culture. It has been considered that a photochemically synthesized compound (substance B) is the essential trigger of sporulation. This essential compound is believed to be formed from a precursor synthesized in the vegetative plasmodium in either light or dark.

$$\text{Plasmodium} \xrightarrow[\text{or dark}]{\text{light}} \text{Substance A} \xrightarrow[\text{light}]{} \text{Substance B} \to \text{Sporulation}$$

It has been suggested that substance A may be produced from a metabolite such as niacin or a metabolite synthesized through reactions catalysed by niacin, that substance B is produced by a photochemical reaction involving substance A and the photoreceptor and that substance B can cause an inactivation of sulphydryl groups which appear to inhibit sporulation (Gray and Alexopoulos, 1968).

The effect of light is not retained when a plasmodium is kept for 10–20 h at 4°C after illumination, nor can conditioned media or extracts from illuminated cultures induce sporulation in starved plasmodia. If the light stimulation is given only to one half of a plasmodium only the illuminated half will sporulate. Thus the effect of light is local and cannot be moved to other distant parts. Such observations may imply that the light effect functions on membrane organization and not primarily at a cytoplasmic level. The metabolic changes accompanying the light effect include an inhibition of respiration, inhibition of glucose uptake, fluctuation of ATP levels and an increase in non-ferric iron. Fluctuations also occur in several oxidase systems.

Since the entire process of sporulation occurs under starvation conditions the energy and building units required for the complex reactions of sporulation, viz. coalescence, stalk and spore formation must be derived from a reorientated metabolism within the plasmodium—a situation already seen in *Dictyostelium discoideum*. Does evidence exist for considering that transcription of the genome and differential protein synthesis are part of the sporulation programme?

The critical point of no return occurs about 3 h after the period of illumination and plasmodia returned to growing media after this point will complete sporulation. Sporangial bodies form approximately 8 h after illumination. Approximately 12 h after the period of illumination mitosis occurs and this is followed almost immediately by morphological changes within the developing sporangia separating the nuclei into mononucleate spores. Meiosis occurs 1–2 days after cleavage. Thus the nuclear cycle does not appear to be in synchrony with the morphological cycle of sporulation. As yet it cannot be determined whether the morphological and nuclear events are triggered at the same or different times. If they are triggered at the same time it may be that the events of the morphological development proceed faster than do the nuclear events.

Unlike spherulation the process of sporulation is sensitive to actinomycin D. Studies with actinomycin D indicate that the mRNA synthesis for sporulation occurs during starvation and illumination and that synthesis is completed about 2 h after the period of illumination since actinomycin D applied after that time is without effect. Although changes in RNA synthesis have been shown to occur during the starvation period they are insufficient to induce sporulation as further RNA synthesis must occur during and after illumination to bring about complete sporulation.

The most important period of RNA synthesis is now considered to occur during illumination and for up to 2–3 h afterwards. These considerations have been strengthened and confirmed by many advanced experimental techniques including the measurement of actinomycin D sensitivity of sporulation during this period, and the increased (^{3}H) uridine and ^{32}P incorporation into RNA. A complete survey of the methods of study has been compiled and discussed by Sauer (1973).

After commitment to sporulation no major RNA synthesis was required for the completion of sporulation. Protein synthesis was essential at all stages of sporulation and some evidence exists for the formation of new proteins during sporulation. The role of light in sporulation is believed to be concerned with stimulating transcription and also aligning the protein-synthesizing systems in a manner permitting the ready translation of new information from the sporulating genome.

Thus sporulation in *Physarum polycephalum* is clearly under transcriptional control and is an example of selective gene activation of a differentiation process.

Summary

Physarum polycephalum can show three distinct and quite separate developmental growth forms depending on the environmental conditions: a cyclic vegetative mitotic phase, a transient vegetative phase of sclerotization or spherulation, and an irreversible sexual phase.

The most remarkable feature of the multinucleate plasmodium is the observation that all nuclei in one plasmodium divide synchronously and that the plasmodium is sufficiently large to permit chemical analysis without the necessity of applying micro-methods.

There is indirect evidence for the stimulation of nuclear division by the production of specific cytoplasmic, diffusible protein factors late in prophase. Since no other eukaryotic organism offers such a unique degree of nuclear synchrony the isolation, characterization and identification of these factors will represent a major advance in understanding the phenomenon of nuclear division in eukaryotic cells. The implications of these studies extend far beyond the confines of mycology and may ultimately lead to a fuller understanding of growth and cancer in higher organisms including man.

Under certain cultural conditions the plasmodia will cease active growth and become encysted. The encystment of microplasmodia presents an excellent experimental system for this form of development because it can be induced easily, consistently and quickly by osmotic shock or by starvation. Although protein synthesis is a necessary requirement of the overall process the synthesis of new RNA does not seem essential. Perhaps this is an example of stable messenger RNA present during the normal vegetative growth becoming fully active when the tight correlation of mitosis and DNA synthesis is altered by the metabolic changes that prelude spherulation. The extensive enzyme studies now being carried out with spherules produced by several types of induction show the care that must be taken when interpreting fluctuating enzyme patterns during a developmental change. Furthermore, such studies also indicate the overwhelming role of the nutrient in determining developmental patterns. For too long studies of development have made use of complex media with which it is impossible to differentiate between a true developmental response or merely a secondary effect of a medium component.

Sporulation is undoubtedly the most complex developmental process exhibited by *Physarum polycephalum* and ultimately leads to genetic recombination and propagation by spores. Two sets of conditions are required for sporulation, one coming from the substrate or medium and the other by illumination. An intricate balance must be maintained during this stage of development and any deviation can stop development. Selective gene activation of differentiation does appear to occur during sporulation. There are few specific biochemical markers for sporulation and only when these have been identified will a full understanding of sporulation be achieved. Again extremely complex media are still being used and the implications of this must be considered before biochemical interpretations can be made.

Thus this simple micro-organism which in nature exists in dark secluded environments may one day assist scientists to provide answers to some of the more significant biological problems of our time.

Recommended Literature

General Reviews

Alexopoulos, J. (1966). Morphogenesis in the Myxomycetes. *In* "The Fungi" **2,** 211–234. Ed. Ainsworth, G. C. and Sussman, A. S. Academic Press, New York and London.

Cummins, J. E. (1969). Nuclear DNA replication and transcription during the cell cycle of *Physarum*. *In* "The Cell Cycle", pp. 141–158. Ed. by Whitson, G. and Cameron, I. Academic Press, New York and London.

Cummins, J. E. and Rusch, H. P. (1968). Natural synchrony in the slime mould *Physarum polycephalum*. *Endeavour* **27,** 124–129.

Cray, W. D. and Alexopoulos, J. (1968). "Biology of Myxomycetes". The Ronald Press Co, New York.

Rusch, H. P. (1970). Some biochemical events in the life-cycle of *Physarum polycephalum*. *In* "Advances in Cell Biology" **1,** 297–327. Ed. Prescott, D. M., Goldstein, L. and McConkey, E. Appleton-Century-Crofts, New York.

Sauer, H. W. (1973). Differentiation in *Physarum polycephalum*. *Symposium Society of General Microbiology* **23,** 375–405.

Genetics

Dee, J. (1966). Multiple alleles and other factors affecting plasmodium formation in the true slime mould *Physarum polycephalum* Schw. *Journal of Protozoology* **13,** 610–616.

Cultural Techniques

Daniel, J. M. and Baldwin, H. (1964). Methods of culture for plasmodial myxomycetes. *In* "Methods in Cell Physiology", pp. 9–41. Ed. Prescott, D. M. Academic Press, New York and London.

Mohberg, J. and Rusch, H. P. (1969). Growth of large plasmodia of Myxomycete *Physarum polycephalum*. *Journal of Bacteriology* **97,** 1411–1418.

The Plasmodium

Braun, C., Mittermayer, C. and Rusch, H. P. (1965). Sequential temporal replication of DNA in *Physarum polycephalum*. *Proceedings of the National Academy of Sciences, U.S.A.* **53,** 924–931.

Chin, B., Briedrich, P. D. and Bernstein, I. A. (1972). Stimulation of mitosis following fusion of plasmodia in the Myxomycete *Physarum polycephalum*. *Journal of General Microbiology* **71**, 93–101.
Hüttermann, A., Porter, M. T. and Rusch, H. P. (1970). Activity of some enzymes in *Physarum polycephalum*. **1**. In the growing plasmodium. *Archiv für Mikrobiologie* **74**, 90–100.
Mittermayer, C., Braun, R. Chayka, T. G. and Rusch, H. P. (1966). Polysome patterns and protein synthesis during the mitotic cycle of *Physarum polycephalum*. *Nature, London* **210**, 1133–1137.
Mohberg, J. and Rusch, H. P. (1971). Isolation and DNA content of nuclei of *Physarum polycephalum*. *Experimental Cell Research* **66**, 305–316.
Rusch, H. P., Sachsenmaier, W., Behrens, K. and Gruter, V. (1966). Synchronization of mitosis by the fusion of the plasmodia of *Physarum polycephalum*. *Journal of Cell Biology* **31**, 204–209.
Zellweger, A. and Braun, R. (1971). RNA of *Physarum*: template replication and transcription in the mitotic cycle. *Experimental Cell Research* **65**, 424–432.

Spherulation

Chet, I. and Rusch, H. P. (1969). Induction of spherule formation in *Physarum polycephalum* by polyols. *Journal of Bacteriology* **100**, 673–678.
Chet, I. and Rusch, H. P. (1970). RNA differences between sphorulating and growing microplasmodia of *Physarum polycephalum* as revealed by sedimentation pattern and DNA–RNA hybridization. *Biochimica et Biophysica Acta* **209**, 559–568
Goodman, E. M., and Rusch, H. P. (1970). Ultrastructural change during spherule formation in *Physarum polycephalum*. *Ultrastructural Research* **30**, 172–183.
Hüttermann, A. (1972). Isoenzyme pattern and *de novo* synthesis of phosphodiesterase during differentiation (spherulation) in *Physarum polycephalum*. *Archiv für Mikrobiologie* **83**, 155–164.
Hüttermann, A., Elsevier, S. M. and Pschrich, W. (1971). Evidence for the *de novo* synthesis of glutamic dehydrogenase during the spherulation of *Physarum polycephalum*. *Archiv für Mikrobiologie* **77**, 74–85.
Hüttermann, A., Porter, M. T. and Rusch, H. P. (1970). Activity of some enzymes in *Physarum polycephalum*. **11**. During spherulation (differentiation). *Archiv für Mikrobiologie* **74**, 283–291.
Jump, J. A. (1954). Studies on sclerotization in *Physarum polycephalum*. *American Journal of Botany* **41**, 561–567.
Kikuchi, M. (1971). Studies on sclerotia of a Myxomycete *Physarum polycephalum*. 1. Electron microscopic observations in the processes of sclerotization and reversion therefrom. *Science Reports of the Tokyo Kyoiku Daigaku* **14**, 215–223.

Sporulation

Daniel, J. W. (1966). Light-induced synchronous sporulation of a Myxomycete: the relation of initial metabolic changes to the establishment of a new cell state. *In* "Cell Synchrony", pp. 117–152. Ed Cameron, I. and Padilla, G. M. Academic Press, New York and London.
Goodman, E. M., Sauer, H. W., Sauer, L. and Rusch, H. P. (1969). Polyphosphate and other phosphorous compounds during growth and differentiation of *Physarum polycephalum*. *Canadian Journal of Microbiology* **15**, 1325–1331.

Jochusch, B., Sauer, H. W., Brown, D. F., Babcock, K. L. and Rusch, H. P. (1970). Differential protein synthesis during sporulation in the slime mold *Physarum polycephalum*. *Journal of Bacteriology* **103,** 356–363.

Lestourgeon, W. M. and Rusch, H. P. (1971). Nuclear acidic protein changes during differentiation in *Physarum polycephalum*. *Science* **174,** 1233–1236.

Losick, R., Shorenstein, R. S. and Sonnenshein, A. L. (1970). Structural alteration of RNA polymerase during sporulation. *Nature, London* **227,** 910–913.

Mohberg, J. and Rusch, H. P. (1970). Nuclear histone in *Physarum polycephalum* during growth and differentiation. *Archives of Biochemistry and Biophysics* **138,** 418–432.

Sauer, H. W., Babcock, K. L., and Rusch, H. P. (1969). Changes in RNA synthesis associated with differentiation (sporulation) in *Physarum polycephalum*. *Biochimica et Biophysica Acta* **195,** 410–421.

Sauer, H. W., Babcock, K. L., and Rusch, H. P. (1969). Sporulation in *Physarum polycephalum*. A model system for studies on differentiation. *Experimental Cell Research* **57,** 319–327.

4 The Fungal Spore

4.1 What is a Spore?

The fungus spore is a distinct part of the mycelium, highly specialized for reproduction, survival and dispersal. In contrast to the vegetative mycelium, the spore is normally delimited from the thallus and is characterized by a minimal metabolic turnover, low water content and lack of cytoplasmic movement (Gregory, 1966). Spores may be produced by asexual (see Chapter 6) or sexual (see Chapter 7) means. Asexual spores are produced by mitotic division and may be homokaryotic or heterokaryotic. Conversely, sexual spores are derived by meiosis and consequently provide new genetic recombinations.

Fundamentally, spores may be characterized as memnospores or xenospores (Table 4.1). Memnospores normally do not move from

TABLE 4.1. Characterization of fungal spores. From Gregory (1966)

Xenospores	Memnospores
Definition:	
Dispersed from place of origin	Remaining at place of origin
Tendencies:	
Completely separating from the mycelium	Often not completely separating from the mycelium
Usually with a definite launching mechanism	Often freed by lysis
Size very various in different species, often minute	Tend to be large
Shape very various	Tend to spherical shape
Survival tends to be short, and the spores often thin-walled	Often very durable and thick-walled
Usually germinating readily in suitable conditions	Often needing a resting period, or only germinating after applying some specific stimulus, shock, or nutrient, or removing an inhibitor

their point of development and conserve the species over periods un-favourable to growth, e.g. drought and starvation. Examples of this type are chlamydospores and zygospores which characteristically have very thick resistant walls. Xenospores become disengaged from the parent thallus and are dispersed actively or passively to more or less distant sites thus ensuring a wider geographical spread for the species. In suitable new environments, such spores may initiate new mycelia directly or mate with existing haploids and initiate new sexual growth. Spores may be hyaline or coloured, one-celled, two-celled, or many-celled, and are normally produced in vast numbers.

Spores are generally considered to be a more or less dormant stage in the life cycle of a fungus. Apart from their ecological role in nature, spores are used in the laboratory for maintenance of cultures and for purposes of inoculation to produce biomass for experimental or industrial needs. Fungal spores also show a widespread and marked ability to transform chemically certain apparently unrelated compounds such as steroids, fatty acids and triglycerides. For these specific transformations, spores can be several times more efficient on a dry weight basis than mycelium. This biocatalytic ability of spores is not yet fully explored (Vezina, Singh and Sinha, 1968).

When a spore encounters a suitable environment, it will germinate. Germination comprises the many processes and changes that occur during the resumption of development or growth of a resting structure and its subsequent transformation to a morphologically different structure. In most fungi this will usually involve the change from a non-polar spore to a polar germ tube which will continue growth by extension at the tip. While germination has been defined by Manners (1966) as the formation of a germ tube from the dormant spore, Sussman (1966) has considered that any measurable irreversible change can be the accepted criterion of germination. Thus spores may be considered to be the beginning and the end of the developmental cycle of a fungus.

Since fungal spore germination involves changes in both metabolism and morphology it provides not only a model system for studying the resumption of growth of a dormant cell but also for studying the basic biochemical and structural events controlling cellular differentiation.

4.2 The Nature of Dormancy

Concepts of Dormancy

Dormancy has been considered to be any rest period or reversible interruption of the phenotypic development of an organism. Since fungal spores are no longer engaged in the synthesis of new cellular material and have a much reduced metabolic activity, they can be considered to be resting cells and to exhibit in some degree the condition of dormancy.

Two types of dormancy are apparent in fungal spores: constitutional and exogenous. In constitutional dormancy, spores will not germinate when placed under favourable environmental conditions but first require an activation process to overcome an innate property of the dormant state. This property may be a barrier to the penetration of nutrients, a metabolic block or the production of a self-inhibitor. Conversely, in exogenous dormancy the spores are inactive because of unfavourable chemical and/or physical conditions and will quickly resume growth if placed in a suitable environment. In this type of dormancy germination is controlled by the need for essential external growth factors, whereas in constitutional dormancy there is a barrier preventing the environmental growth factors initiating the process of germination.

Dormancy is widespread in nature and is presumably of considerable selective advantage during the course of evolution.

The Breaking of Dormancy

Whereas it is possible to overcome exogenous dormancy by returning the spores to a suitable growth environment, constitutional dormancy can only be overcome by selective treatments that are not normally required for vegetative growth.

Although many types of fungal spores exhibit constitutional dormancy the ascospore of *Neurospora crassa* has undoubtedly received the greatest experimental attention. The ascospore has at least 5 distinct wall layers, one of which contains melanin, which undoubtedly contribute to its impermeability and heat and radiation resistance (Sussman, 1966). The ascospore relies entirely on endogenous substrates during the dormant condition and after activation requires only water for germination and outgrowth.

For the *Neurospora crassa* ascospores, heat shock is essential and a temperature of 50–60°C for 20 min. will ensure breakage of dormancy. Treatment with furfural, heterocyclic compounds and certain other chemical compounds can also be successfully used. In some cases, e.g. the teliospores of *Puccinia graminis*, cold treatment is essential. This normally occurs with fungi that are pathogenic on plants and probably ensures that the spore will not germinate until after the winter period. Heat resistance in *N. crassa* ascospores is due to the exospore wall since removal of this wall layer is sufficient to permit some germination at much lower temperatures than are normally required.

Dormancy is often maintained by germination inhibitors. Such compounds may have been formed by the parent mycelium and carried with the spore or they may be formed directly by the spore (Allen, 1965). The effects of these inhibitors can be overcome by washing the spores in water or by the addition of chemicals that will counteract the inhibitor. Spore self-inhibitors are readily observed in large populations of uredospores of rust fungi but have also been reported for oospores and conidia of downy mildews and for the conidia of many saprophytic fungi. Germination stimulants such as coumarins and phenols have been isolated from some spores and it has been suggested that a balance between inhibitors and stimulants may be involved in the *in vivo* regulation of germination.

An interesting example of stimulation of spore germination is shown by the Basidiomycete *Agaricus bisporus*. For many years it has been known that germination of the basidiospore is promoted by *A. bisporus* hyphae growing near by. The effect is shown to be caused by a volatile agent which could also be obtained from yeast and other fungi. Isovaleric acid and isoamyl alcohol have been identified as metabolites of *A. bisporus* and are considered to be involved in the volatile stimulation of basidiospore germination. This "triggering" reaction of isovalerate probably consists of the removal of a CO_2-self-inhibitor by participation of β-methylcrotonyl-CoA carboxylase (Rast and Stauble, 1970).

Chemical Changes Associated with Dormancy

It is considered that many, if not most, fungal spores require the simultaneous utilization of lipids and carbohydrates for germination to proceed.

Many chemical studies have been made on the *Neurospora* ascospore activation process and it is clear that glycolysis occupies an important and perhaps controlling influence. There is no evidence for the existence of the HMP pathway in ascospores, and glycolysis may be the main pathway for glucose degradation and energy production during the early stages of activation. The full complement of TCA cycle acids is not found until approximately one hour after activation although all the enzymes of the pathway appear to be present. Evidence has been presented that mRNA is not stored in dormant ascospores of *N. crassa* and that they do not contain polyribosomes although ribosomes and tRNA are present. Presumably germination is accompanied by mRNA synthesis and subsequent polyribosome formation.

Whereas dormant ascospores metabolize lipids as an energy source, activated spores also use trehalose which disappears rapidly upon activation. The low level of all metabolites actually found in most dormant fungal spores probably results from a metabolic block or nutritional deficiency that must be removed before germination can occur. Trehalase, which hydrolyses trehalose to glucose, has been detected in dormant and activated ascospores. There have been numerous hypotheses presented to explain the inability of dormant ascospores to use trehalose: (a) the *de novo* synthesis of a trehalose degrading enzyme during activation; (b) the spatial separation of trehalose and trehalase in the dormant spore and the subsequent break-down of this compartmentalization during activation; (c) the removal of an inhibitor of trehalase during activation; (d) the conversion of a precursor into the enzyme trehalase and (e) a series of interlocking enzymatic reactions are shifted from one steady state to another (Eilers *et al.*, 1970). Since the enzyme trehalase can be detected in dormant spores, (a) is unlikely and the change in activity is too rapid to consider *de novo* enzyme synthesis.

During induction of germination with either heat shock or furfural treatment, an examination of a variety of metabolic intermediates showed rapid and significant changes in those involved in glycolysis. Associated with a decrease in trehalose concentration, there was a con-comitant increase in the concentration of glucose, G-6-P, F-6-P, PEP, pyruvate, ethanol, malate, OAA, ATP, NADPH and NADH (Eilers *et al.*, 1970). These rapid metabolic changes are most probably due to enzymatic activation either by method (c) or (e).

Similar results have been achieved with dormant species of *Phycomyces blakesleeanus*. After heat shock there was a rapid increase in respiration and in the concentration of pyruvic acid, acetaldehyde and ethanol. Trehalose concentration also rapidly decreased at activation and there was a parallel stimulation of glycolysis (Rudolph and Ochsen, 1969).

In certain species, dormancy can be broken and germination initiated by the addition of catalytic amounts of specific substances. Studies with dormant spores of *Penicillium roquefortii* would suggest that induction of the tricarboxylic acid cycle may be involved in the activation of the spores and that ATP may be the trigger. Dormant spores of *P. roquefortii* can only oxidize octanoic acid slowly. The addition of small amounts of sugars or amino acids not only breaks dormancy but also causes a large increase in the oxidation of octanoic acid. Thus the termination of dormancy in this fungus is an energy dependent process. It has been suggested that the activation of many fungal spores may require the supply of key intermediates such as acetyl CoA for the triggering of dormant energy-producing systems and that this activation is brought about because specific stimulatory compounds can readily penetrate the membrane of the cell (Lawrence and Bailey, 1969).

4.3 Spore Germination

When all barriers of dormancy have been removed, the spores require a favourable moisture level and temperature together with certain substances from the environment before they can commence the germination process. Aerobic respiration appears to be a normal requirement for most spores since it is doubtful if any spore can initiate germination under conditions of true anaerobiosis. Many spores appear to require carbon dioxide, and fixation of CO_2 has been reported in many fungi. Some spores with adequate organic reserves appear to need only water for germination although the majority of spores tested appear to have a requirement for organic carbon at some stage in their germination process (Fig. 4.1).

When all of these conditions have been met, the spore is now capable of germination. The germination process can be conveniently divided into three main phases (Manners, 1966):

(1) The internal physiological and morphological changes that will occur within the spore wall before any outgrowth occurs. Such changes can only be monitored by chemical analysis and by electron microscopy.
(2) The act of protrusion of the germ-tube from the spore wall.
(3) The elongation of the germ-tube and the establishment of polar growth.

Changes in Morphology and Fine Structure (non-motile spores)

Spherical Wall Growth

With the exception of some powdery mildews, most spores will undergo some degree of swelling during germination prior to germ-tube emergence. Spores which require nutrients for germination normally undergo more swelling than spores that germinate in water alone. With *Aspergillus niger*, Yanagita (1957) distinguished two phases within the swelling process. An initial or endogenous swelling occurred in deionized water due probably to physical stretching caused by imbibition and was independent of nutrients. A second or exogenous swelling was nutrient dependent and undoubtedly indicates a period of active metabolism and growth. The term spherical growth, devised by Bartnicki-Garcia *et al.* (1968) to describe the swelling phase of *Mucor rouxii* sporangiospores, has now been generally accepted.

The amount of spherical growth that can occur during germination

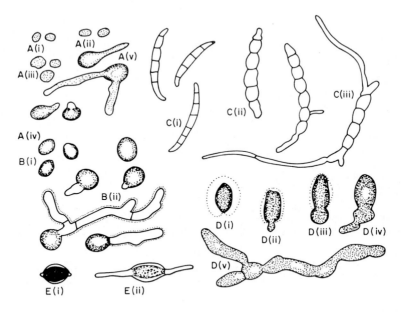

Fig. 4.1. Germinating spores, all approximately × 500. A, *Rhizopus arrhizus*, sporangio-spores. (i) unsoaked spores, prolate spheroids with longitudinally striate walls; (ii) and (iii) spores soaked in glucose-salts medium at 30°C for 2 and 4 h respectively, showing progressive swelling, more rapid in direction of short axis, giving almost spherical shape; (iv) and (v) soaked for 5 and 6 h respectively, germination has taken place but swelling continues. B, *Botrytis cinerea*, conidia. (i) unsoaked spores; (ii) spores soaked in water at 25°C for 4 h; swelling has occurred but the increase in volume is relatively less than in *R. arrhizus*; note slight constriction of germ-tube at point of emergence, and mucilage sheathing younger parts (shown as dotted line). At base of older germ-tubes the mucilage has consolidated. C, *Fusarium lateritium* Nees conidia. (i) unsoaked sickle-shaped spores; (ii) spore soaked in water at 25°C for 3 h; note swelling of individual cells and consequent alteration in outline of spore; (iii) germinating spores, after 5 h soaking; any cell is potentially capable of producing a germ-tube but the first germ-tubes usually emerge from the end cells of the spore; note constriction at base of germ-tube. D, *Sordaria fimicola* (Roberge) Cesati and de Notaris, ascospores. (i) unsoaked spore, dotted line indicates gelatinous sheath; (ii) spore soaked 3 h in 2% malt extract at 25°C; no significant swelling of spore; note shrinkage of sheath and emergence of small vesicle; (iii)–(v) stages in germination, showing enlargement of vesicle and development from it of one or more germ-tubes, note shrinkage of spore and development of vacuole in vesicle. E, *Melanospora zamiae* Corda, ascospores. (i) unsoaked spore, note polar papillae and spore packed with oil drops; (ii) germinated spore with germ-tube emerging at each pole, note alteration of spore shape, disappearance of all but a few oil drops and slight vesicular swelling at base of each germ-tube. From Hawker (1966).

varies with the species and in most spores leads to an approximate 2–3 fold increase in the diameter of the spore (Fletcher, 1969). Germ-tube emission normally occurs before maximum spherical growth has been achieved. An increase in the incubation temperature to a level above that normally used in spore germination studies (25–30°C) has been found to have marked effects on the morphology of germination of *Aspergillus niger* spores (Anderson and Smith, 1972). As the temperature of incubation is increased above 35°C there is an increase in the extent of spherical growth and a reduction of germ-tube outgrowth. When the temperature of incubation is held at 44°C for 48 h the spores become large spherical cells which can no longer produce a germ-tube. Dormant spores of *A. niger* are approximately 3·5 μm in diameter and during normal germination will achieve a maximum diameter of 6–7 μm before germ-tube outgrowth. Spores that have been treated for 48 h at 44°C reach a final diameter of 20–25 μm. No spherical growth occurs when the spores are incubated at temperatures above 45°C. Changes in spore diameter and spore volume in relation to temperature are shown in Table 4.2 and Fig. 4.2 and 4.3. Subsequent incubation of these giant

TABLE 4.2. Comparison of spore diameter, spore surface, and spore volume of unswollen spores and large SG spores of *Aspergillus niger* produced after 48 h cultivation at 44°

	Spore diameter μ	Spore surface area, μ^2	Spore volume μ^3
A			
Unswollen spores	3·5	38·5	22·4
B			
Large SG spores	19·6	1207	3933
Ratio B/A	5·6	31·4	175·6

cells at 30°C leads to the direct outgrowth of conidiophores (Anderson and Smith, 1971). This microcycle conidiation will be discussed in Chapter 6.

In order to understand the increase in spore size during germination it is necessary to examine first the nature of the spore wall before the onset of germination. There have been few studies comparing the chemistry of wall structure of spores and of vegetative cells. In *Mucor rouxii*, the polysaccharides of vegetative walls are primarily glucosamine polymers (chitosan and chitin) and polyuronides with the conspicuous absence of glucose polymers. In the walls of the sporangio spores the principal component is a glucose polymer; glucosamine is present in much smaller amounts and in combination with protein and melanins. Polyuronides are rarely found. It is probable that the changes in wall chemistry give the spore a much greater survival value.

Although there have been many studies on the ultrastructure of dormant spores, few have taken the precaution to use dry harvested spores. The use of spores collected by aqueous methods has exposed

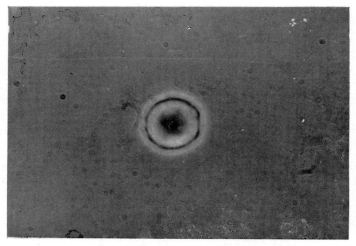

Fig. 4.2. Conidium of *Aspergillus niger* at time of inoculation of culture medium. Phase contrast × 1840.

them to rehydration even if only for a brief period, and has probably led to the confusion of the origin of the germ-tube outgrowth.

Using dry harvested spores of *Aspergillus nidulans*, Florance *et al.* (1972) have made a careful examination of the spores during germination. Initially, the cell wall of the mature dormant spore consists of three layers (Fig. 4.4) with the outer pigmented layer composed of a network of rodlets (Fig. 4.5). Shortly after hydration, two new layers appear in the spore wall (Fig. 4.6). They have suggested that these new layers do not arise by *de novo* synthesis of new wall material but rather are derived from the innermost wall layer by imbibition and molecular reorientation.

Bartnicki-Garcia (1973) has proposed that a new layer (the vegetative wall) is synthesized *de novo* under the spore wall, and by a process of wall synthesis not unlike that which occurs in a yeast cell leads to

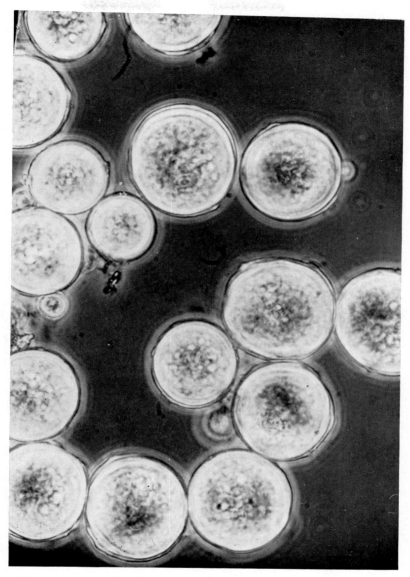

Fig. 4.3. Large spherical cells of *Aspergillus niger*. Conidia were incubated in the culture medium at 44°C for 48 h. Conidia increased in size by spherical growth and did not produce germ-tubes. Phase contrast ×690.

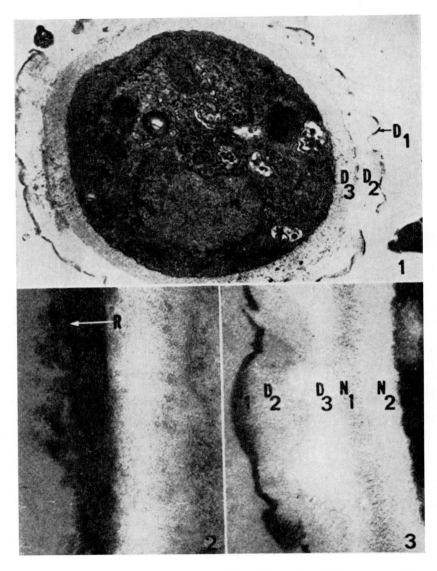

Fig. 4.4–4.6. *Aspergillus nidulans* conidia. (1). This section of dormant conidium, showing ER, nucleus, concentric membranous structure (CMS), vacuoles (V), vesicles (Ve), mitochondria, ribosomes, and three-layered wall (D_1, D_2 and D_3), ×35, 360. (2). Portion of conidium wall hydrated 10·1/2 h. Note rodlets (R) in outer (D_1) layer, ×108 160. (3). Portion of conidium wall hydrated 30 min. Note new layers (N_1 and N_2) ×86 580. From Florance, Denison and Allen (1972).

uniform deposition of wall material. Studies with (^{3}H) N-acetyl-D-glucosamine have confirmed this hypothesis and clearly show spherical growth (Fig. 4.7). It is this non-polarized pattern of wall growth with a uniform isotropic deposition of cell wall polymers that takes place during the spherical growth phase of spore germination. During the spherical growth phase, not only is there an increase in the size of the

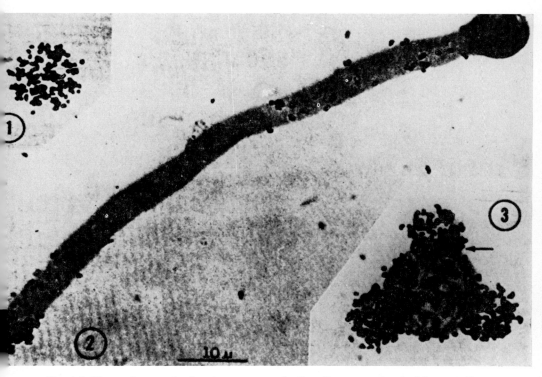

Fig. 4.7. Photomicrographs showing patterns of cell-wall construction in *Mucor rouxii*. (1). Shows germinated sporangiospore prior to germ-tube emission with disperse pattern of wall synthesis. (2). Shows a hypha with apical pattern. (3). Depicts a yeast cell with three buds showing disperse patterns; one of the buds (arrow) also exhibits a band of basal wall synthesis probably related to septum formation. Cells were grown anaerobically under nitrogen (1 and 2) or 30% carbon dioxide. From Bartnicki-Garcia (1969).

cell but there is also a considerable increase in the dry weight which is linked to an essential requirement for exogenous nutrients. In the giant cells of *Aspergillus niger*, although there is a 30-fold increase of surface area the spore walls remain thick.

Thus an essential feature of most germinating spores is the existence of a period of limited spherical growth in which new wall material is deposited within the original spore coat.

Germ Tube Outgrowth

Before a germ tube protrudes from the spore coat, the pattern of wall synthesis must become gradually polarized with most of the wall synthesis taking place at one or a few areas of the spherical cell surface. These areas subsequently become the apical dome of the emerging germ tube and the continuous polarized production of cell wall material leads to the formation of a hyphal tube (see Chapter 5).

It is now considered that the new hyphal wall is continuous with the innermost layer of the spore wall, a layer not apparent in the dormant condition but appearing after hydration. The key factor responsible for

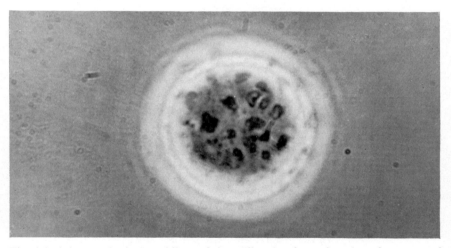

Fig. 4.8. A large spherical conidium of *Aspergillus niger* focused to show fragments of the dark pigmented external coat adhering to the expanding spore wall. Phase contrast ×2000.

the switch from spherical growth to germ-tube emission is probably that which causes cell wall formation to become polarized. Again, in the case of the giant cells of *Aspergillus niger*, the effect of prolonged incubation at 44°C is in some way to prevent the normal polarization from occurring. This must obviously raise the question as to how large spherical cells could become if the spherical growth process could proceed in the absence of polarization.

The mechanism whereby the germ-tube passes through the spore coat is not well understood. In some fungal spores there are predetermined thin areas called germ-pores through which the germ tubes emerge (Fig. 4.1). However, not all spores have these pores, and outgrowth is achieved by rupturing of the outer wall. Mechanical rupturing as well as degradation by hydrolytic and/or proteolytic enzymes have been suggested. The remains of the original spore coat can readily be seen on the giant cells of *A. niger* (Fig. 4.8).

Thus the outer spore wall may now be considered to be a transient coat destined exclusively for the protection of the protoplast and perhaps of the early stages of germination. Once germination has commenced, it is replaced by a chemically different cell wall, the vegetative wall, which is formed underneath the spore wall and after a variable period of spherical growth becomes polarized and develops as the new hyphal tube.

ULTRASTRUCTURE OF GERMINATION

A general feature of most dormant spores is a sparse endoplasmic reticulum which rapidly increases during germination. Although some mitochondria are present in the dormant condition it is considered that they may not be fully functional. The general decrease in mitochondrial size which occurs during germination together with the concurrent increase in numbers suggests that the original mitochondria multiply by fission. Ungerminated sporangiospores of *Phycomyces* contain the complete cytochrome system although the mitochondria are poorly developed. The increase in cytochrome content parallels the increase in mitochondrial numbers (Keyhani *et al.*, 1972). It is probable that the synthesis of cytochrome is correlated with the conversion of promitochondria to normal mitochondria containing fully developed cristae. Germination of all spores is paralleled by a marked increase in respiration.

The vesicular hypothesis for hyphal tip growth proposes that exocytotic vesicles produced by the endomembrane system of the cell accumulate at the site of growth, contribute to the expanding apex and thus mediate cell extension (for full details see Chapter 5).

Most early electron microscope studies failed to observe apical vesicles at the site of spore germination. However, the recent studies by Bracker (1971) and Grove (1972) with several fungi have conclusively shown that germ-tube apices contain cytoplasmic vesicles similar to the secretory vesicles formed at the tips of vegetative hyphae. Such vesicles are present at the earliest stage of development of the outgrowth and remain up to the establishment of branched hyphae. The number of vesicles in germ-tube tips is generally less than that found in actively growing hyphal tips (Fig. 4.9).

The failure of most previous studies to detect apical vesicles at the site of spore germination was most probably due to the poor prefixation

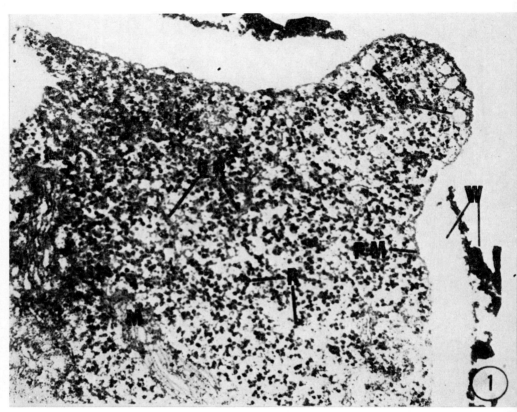

Fig. 4.9. Formation of vesicles during germ-tube formation in *Aspergillus parasiticus*. Portion of a germinating conidium showing a germ-tube emerging through a rupture in the double-layered spore wall (W). Note the cytoplasmic vesicles (V) next to the plasma membrane (PM) in the apex of the germ-tube. Ribosomes (R), mitochondria (M) and endoplasmic reticulum (ER) are shown in the spore cytoplasm, $\times 40\ 000$. From Grove (1972).

treatment. It is now known that any prefixation treatment that stops active growth will immediately lead to the loss of the vesicles. Such studies further demonstrate the delicate nature of growing tips of fungal germ tubes and hyphae. Subcellular apical organization is a very labile property easily disrupted by adverse environmental influences.

In germinating sporangiospores of *Mucor rouxii* a single electron dense organelle (0·2 μm), the apical corpuscle, has been seen to be intimately associated with the apical wall of the incipient germ tube (Bartnicki-Garcia *et al.*, 1968). It has been suggested that this organelle has a role in germ-tube emission and in controlling apical growth. However, in the light of the recent observations of Bracker, there is now some doubt as to the exact nature and role of this organelle.

Motile Spores

THE GERMINATION PROCESS

Studies on the germination process of the motile spore or zoospore has until recently considerably lagged behind that of the non-motile spore. The main reason has been the difficulty of obtaining synchronous germination of large numbers of zoospores. However, methods have recently been developed which allow rapid, complete and moderately synchronous cell transformations (Soll *et al.*, 1969; Cantino, 1971). Most studies of zoospore germination have been made with *Blastocladiella emersonii*. The zoospore of this species is about 7 × 9 μm in size and propels itself by way of a single posterior, whiplash flagellum. There is no definite cell wall but a single, continuous unit membrane which delimits the protoplast. The germination process can take place immediately on release from the sporangium but normally the zoospore moves about for a variable period of time before settling down and initiating germination.

The first major developmental step in the germination process is the settling down of the spore and its conversion into an essentially spherical, cystlike cell (Fig. 4.10). During this phase of development, the flagellum is retracted and there is a marked increase in respiration. This phase is extremely rapid and can be measured in seconds or minutes. The initial conversion of zoospores to spherical cells is markedly influenced by the ionic strength of the environment. Only certain ions are effective (Soll and Sonneborn, 1972). The important feature of encystment is the ensheathing of the protoplast with a rigid cell wall. After a period of spherical growth, a small germ-tube emerges from the cell which will later grow into the branched rhizoidal system. The spherical cell continues to grow and eventually forms the large multinucleate, coenocytic thallus (for details see Chapter 6). Thus, as in zoospore germination, there is an obligate formation of a spherical cell prior to germ-tube emergence.

Fine Structural Changes

The zoospore has no distinct cell wall and contains a highly ordered array of cytoplasmic organelles (Fig. 4.10). Some of the major components are: (*a*) a single posteriorly situated flagellum; (*b*) a nucleus; (*c*) a membrane-bound nuclear cap overlying the nucleus and containing most of the cell's ribosomes; (*d*) a giant mitochondrion positioned eccentrically around the nucleus; (*e*) a membrane-bound "lipid-granule" containing body extending along the outer margin of the long

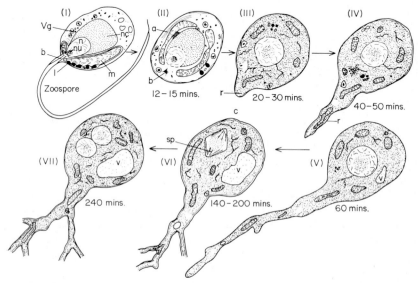

Fig. 4.10. Stages in zoospore germination and early growth of *Blastocladiella emersonii*. A series of diagrammatic sketches to illustrate the morphological and intracellular changes during the first 4 h in mass germination cultures: axoneme of retracted flagellum (a); basal body (b); centriole (c); flagellum (f); lipid (l); mitochondrion (m); nuclear cap (nc); nucleolus (nu); nucleus (n); rhizoid (r); spindle (sp); vacuole (v); vesicle-enclosed granule (vg). From Lovett (1968).

arm of the mitochondrion; and (*f*) gamma particles consisting of a unit membrane enclosing an ellipsoid, bowl-shaped matrix about 0·5 μm in length.

During the formation of the spherical cell the flagellar axoneme is retracted, and localized channels to the plasma membrane for intracellular vesicles associated with the synthesis of the initial cell wall make their appearance. At a later stage, there is breakdown of the nuclear cap and the release of ribosomes into the cytoplasm. The mitochondrion becomes fragmented and numerous small mitochondria can be observed. Associated with these changes there is an increase in volume. The final act of the germination process is the formation of the germ tube. For a full understanding of these changes the authoritative review by Truesdell and Cantino (1971) should be consulted.

Gamma particles have been implicated in the developmental control of germination. Like mitochondria and chloroplasts, they have been shown to contain DNA, and the possibility of their determining their own duplication is now being examined (Myers and Cantino, 1971). It has been suggested that the gamma particles may be involved in phenotypic development. As will be shown in Chapter 6, the germinating cyst of *Blastocladiella emersonii* is totipotent and can produce four phenotypes depending on the environmental conditions: ordinary colourless (OC), late colourless (LC), orange (O) and resistant sporangia (RS). The zoospores from O, OC and LC plants average 8, 12 and 16 gamma particles/cell respectively. The gamma particles appear also to be the main cellular site for chitin synthetase.

Ultrastructural aspects of encystation and cyst germination of the zoospores of *Phytophthora parasitica* have also been studied (Hemmes and Hohl, 1971).

Macromolecular Synthesis during Germination

An important problem in fungal morphogenesis is that of defining the metabolic events which occur during the conversion of a dormant spore into a growing organism. During the dormant period, most spores exhibit a minimal metabolic turnover of structural and informational macromolecules. However, at the onset of germination profound changes occur in metabolism, in particular the initiation and rapid increase in the synthesis of protein, RNA and DNA. The synthesis of these important macromolecules is obviously interrelated and for many years there has existed considerable controversy as to the nature and extent of macromolecular synthesis in the dormant spores, and how this influenced or was influenced by germination.

One way to study the regulatory system involved in protein and nucleic acid biosynthesis during germination would be to determine whether or not the components of the protein synthesizing apparatus are present and functional in the ungerminated spore when compared to the germinated spore.

To this end Van Etten (1969) set out to ask several important questions and to find out how much experimental work had been done and *could* be done to explain the changes in macromolecular synthesis that

occur during spore germination. Since these questions have an importance to basic differentiation in addition to their relevance to spore
germination, they have been restated here.

(1) Are the physical properties and biological activities of ribosomes
from ungerminated spores similar to those of the ribosomes from
germinated spores?

(2) Are ungerminated spores able to conserve mRNA in a stable,
readily translatable form, or is it the lack of mRNA in ungerminated spores which controls protein synthesis?

(3) Are the tRNA molecules for the various amino acids present and
functional in ungerminated spores? Furthermore, are the tRNA
molecules identical for a specific amino acid in germinated and
ungerminated spores?

(4) Are the amino-acyl-tRNA synthetases and transfer enzymes present and functional in ungerminated spores?

(5) Is the spore DNA completely repressed so that even if all of the
components necessary for synthesis of RNA and protein are present and active in ungerminated spores the spore DNA cannot be
transcribed?

(6) Does the ungerminated spore contain RNA polymerase in an
active form so that the various molecular species of RNA can be
synthesized, provided that the DNA is in a form to be transcribed?

(7) Do fungal spore ribosomes lack initiation factors which are
necessary to initiate the translation process of natural mRNA's?

(8) Does the ungerminated spore contain a pool of free amino acids
which is accessible to the protein synthesizing apparatus of the
ungerminated spore?

(9) Are all of the components necessary for synthesizing RNA and
protein present in active forms in ungerminated spores, yet
spatially separated so that they are unable to interact with one
another?

To what extent these questions have been experimentally answered
will be discussed in the remainder of the chapter.

Non-Motile Spores

Studies with several fungi, and in particular *Botryodiplodia theobromae*
have clearly shown that ungerminated spores do contain many if not
all of the components of the protein synthesizing apparatus, viz. ribo-

somes, tRNA, amino-acyl-tRNA synthetases and transfer enzymes, and that these components are biologically active when assayed *in vitro* with synthetic polyuridylic acid. The low level of protein synthesis in spores and its subsequent activation during germination is probably not so much due to the lack of any protein-synthesizing component but rather to a subtle change in some component, such as a change in the iso-accepting species of tRNA.

Several studies using metabolic antagonists, such as cycloheximide and actinomycin D have shown that protein synthesis is an essential requirement for germ-tube outgrowth, whereas RNA synthesis is not (Hollomon, 1969; Brambl and Van Etten, 1970). A comparison of the incorporation kinetics of ^{14}C-leucine increased sharply late in the germination sequence when an actinomycin D sensitive ^{14}C uracil incorporation began. Thus the protein which is involved in the initial germ-tube outgrowth appears to be synthesized on a template of stable mRNA already present in dormant conidia. Alteration of this RNA in such a way that it is able to act as a template for protein synthesis could well be the first step in the germination process.

Density gradient analyses of spore extracts clearly indicate that polyribosomes exist during all the germination stages of some fungi. Polyribosome-enriched fractions isolated from both germinated and ungerminated spores also show the capacity to stimulate amino acid incorporation.

Thus ungerminated spores contain mRNA and this mRNA appears to be immediately functional upon initiation of germination. Furthermore, this mRNA apparently contains enough information to allow the formation of the germ-tube. Actinomycin D-sensitive mRNA synthesis is initiated in the germ-tube.

There are several studies on protein and nucleic acid synthesis in ungerminated and germinated spores which disagree in some way with these conclusions (Van Etten, 1969). However, in many of these cases, spores were harvested in water and it is possible that some steps leading to germination had been activated during the actual harvesting of the spores.

Recently the enzymes involved in RNA synthesis during spore germination have been examined in *Rhizopus stolonifer* (Gong and Van Etten, 1972). These authors have shown quantitative and qualitative changes in RNA polymerase associated with the germination of *R. stolonifer* spores. Germinated spores contained at least three soluble DNA-dependent RNA polymerases whereas ungerminated spores only contained two RNA polymerases: I and III. All three RNA polymerases were inhibited by actinomycin D (Table 4.3).

TABLE 4.3. The effect of antibiotics on the activity of the RNA polymerases I, II and III from germinated spores of *Rhizopus stolonifer*. From Gong and Van Etten (1972)

Antibiotics	Concentration µg/ml	Percentage of Control		
		I	II	III
None (control)	—	100	100	100
Actinomycin D	20	24	34	32
Rifamycin SV	100	100	98	104
Rifampicin	100	92	91	88
α-Amanitin	10	100	73	100
Cycloheximide	50	101	92	103
Chloramphenicol	50	100	100	98

Motile Spores

Almost all studies on macromolecular synthesis in motile spores have been carried out with the zoospores of *Blastocladiella emersonii*. It is interesting to note that almost all of the cell's ribosomes are tightly packaged in the "nuclear cap" surrounding the zoospore nucleus and this cap does not become disorganized until after the zoospore has achieved the initial rounding up prior to germ-tube outgrowth.

There is no detectable synthesis of protein, RNA and DNA during the motile period of the zoospore. The fact that the ribosomes are separated from the main cellular cytoplasm does not constitute the main reason for this inactivity. Rather, it is due to the presence of a ribosome inhibitor which reversibly binds to the nuclear cap ribosomes and in this way blocks their overall function (Schmoyer and Lovett, 1969). The mechanism of action is not yet known.

With the exception of flagellar absorption and germ-tube outgrowth, most other morphological aspects of germination are insensitive to cycloheximide inhibition. Measurable amino acid incorporation only takes place after conversion to the spherical phase, formation of the cell wall and elongation of the mitochondrion. During the period of amino acid incorporation ribosomes are released from the nuclear cap and development of multiple mitochondrial profiles and disappearance of the flagellar axenome and the gamma particles occurs. It has been suggested that the lack of cycloheximide sensitivity could well be due to a lack of cellular penetration. However, there is no doubt that the cycloheximide can enter the cell (Soll *et al.*, 1971). Furthermore, endogenous amino acids are not utilized for early protein synthesis.

An even more remarkable observation is the inability of actinomycin

D to affect the developmental pattern of germination. That the inhibitor molecule could enter the cell was evidenced by the inhibition of ^3H uridine incorporation. The implication from the inhibitor and chemical studies is that much of the structural changes accompanying zoospore germination do not require concurrent RNA and protein synthesis. Although some of the later events of germination, e.g. disappearance of the flagellar axenome and germ-tube emergence, do appear to require *de novo* protein synthesis there is no evidence of new RNA synthesis. Thus the proteins required for the initial germination of the zoospores are made from preformed messages, i.e. long-lived messenger RNA. A parallel situation has been shown to exist with sea-urchin fertilization (see Soll *et al.*, 1971). It has been considered that the protein synthesis for germination must occur during spore formation since zoospores can be made to germinate immediately on release from the sporangium (Soll *et al.*, 1971). It follows then that the events of germination involve mainly reorganization of pre-existing structures rather than *de novo* synthesis.

Clearly the spores of fungi represent an ideal system for investigating the nature of long-lived messenger RNA and also for this new phenomenon of protein reorganization.

Summary

The fungus spore is highly specialized for reproduction, survival and dispersal, and is characterized by low water content, minimal metabolic turnover and lack of cytoplasmic movement. The spore may resume active growth by the process of germination. Most spores appear to have an initial period of non-polarized wall growth leading to an increase in spore diameter. In *Aspergillus niger* and in certain other fungi, non-polarized growth of spores can be continued for long periods resulting in the formation of giant spherical cells. Elevated temperature and above normal concentrations of CO_2 are required to initiate this novel form of vegetative growth. During normal germination the spherical, non-polarized growth phase is of limited duration and at the onset of growth polarization a cylindrical outgrowth, the germ tube, emerges from the spore surface. Associated with the establishment of the germ tube there is the occurrence of vesicles at the point of outgrowth. These vesicles may be involved in transporting the enzymes and substrates required to establish polarized growth. A fuller understanding of the conditions that lead to polarized wall growth (apical tip growth, see Chapter 5) may well come from studies on spore germination. What factors cause spherical non-polarized growth to be replaced by limited polarized growth? Why do the *A. niger* spores continue non-polarized growth for so long when exposed to high temperatures and increased CO_2 concentrations? Is it possible that the spore is initially anaerobic and this prevents the establishment of the factors that control polarized growth? As non-polarized growth continues the penetration of the spore wall by oxygen could well result in an increased aerobic metabolism which may in turn lead to polarization of wall growth. An understanding of the factors initiating and controlling polarization is of major importance in unravelling the complex machinery of spore germination. During germination the new hyphal wall is continuous with the innermost layer of the spore wall, a layer that is not apparent in the dormant condition but appears after hydration. The outer spore wall can be considered as a transient coat destined exclusively for the protection of the protoplast and early stages of germination.

During the dormant period, most spores exhibit a minimal metabolic turnover of structural and informational macromolecules. With the onset of germination there is rapid increase in the synthesis of protein, RNA and DNA. Do ungerminated spores contain a stable, readily translatable form of mRNA? Apparently many ungerminated spores do contain mRNA, and this mRNA can be immediately functional upon initiation of germination. This mRNA is sufficient to permit the early stages of germ-tube development. In *Blastocladiella* the events of zoospore germination appear to mainly involve a reorganization of preexisting structures rather than *de novo* synthesis.

The fungal spore provides a model system for studying the resumption of growth of a dormant cell as well as for studying the basic biochemical and structural events controlling cellular differentiation.

Recommended Literature

General Reviews

Allen, P. J. (1965). Metabolic aspects of spore germination in fungi. *Annual Review of Phytopathology* **3,** 313–334.

Cantino, E. C., Truesdell, L. C. and Shaw, D. S. (1968). Life history of the motile spore of *Blastocladiella emersonii*: a study in cell differentiation. *Journal of Elisha Mitchell Science Society* **84,** 125–146.

Gregory, P. H. (1966). The fungus spore: what it is and what it does. *In* "The Fungus Spore". Ed. Madelin, M. F. Butterworth Science Publications, London.

Madelin, M. F. (Ed.) (1966). "The Fungus Spore". Butterworth Science Publications. London.

Sussman, A. S. and Halvorson, H. O. (1966). "Spores. Their Dormancy and Germination". Academic Press, New York and London.

Truesdell, L. C. and Cantino, E. C. (1971). The induction and early events of germination in the zoospore of *Blastocladiella emersonii*. *Current Topics in Developmental Biology* **6**, 1–44.

Vezina, C., Segal, S. N. and Sinha, K. (1968). Transformation of organic compounds by fungal spores. *Advances in Applied Microbiology* **10**, 221–267.

Dormancy

Eilers, F. I., Ikuma, H. and Sussman, A. S. (1970). Changes in metabolic intermediates during activation of *Neurospora* ascospores. *Canadian Journal of Botany* **16**, 1351–1356.

Keyhani, J., Keyhani, E. and Goodgal, S. H. (1972). Studies on the cytochrome content of *Phycomyces* spores during germination. *European Journal of Biochemistry* **27**, 527–534.

Lawrence, R. C. and Bailey, R. W. (1969). Evidence for the role of the citric acid cycle in the activation of spores of *Penicillium roquefortit*. *Biochimica ei Biophysica Acta* **208**, 77–86.

Rast, D. and Stauble, E. J. (1970). On the mode of action of isovaleric acid in stimulating the germination of *Agaricus bisporus* spores. *New Phytologist* **69**, 557–566.

Rudolph, H. and Ochsen, B. (1969). Trehalose-Umsatz warmeaktivierter Sporen von *Phycomyces blakesleeanus*. *Archiv für Mikrobiologie*, 65, 163–171.

Sussman, A. S. (1966). Dormancy and spore germination. *In* "The Fungi" **2**, 733–764. Ed. Ainsworth G. C. and Sussman, A. S. Academic Press, New York and London.

Sussman, A. S. (1969). The prevalence and role of dormancy. *In* "The Bacterial Spore", pp. 1–33. Ed. Gould, G. W. and Hurst, A. Academic Press, London and New York.

Changes in Morphology and Fine Structure

Anderson, J. G. and Smith, J. E. (1971). The production of conidiophores and conidia by newly germinated conidia of *Aspergillus niger* (microcycle conidiation). *Journal of General Microbiology* **69**, 187–197.

Anderson, J. G. and Smith, J. E. (1972). The effects of elevated temperature on spore swelling and germination in *Aspergillus niger*. *Canadian Journal of Microbiology* **18**, 289–297.

Bartnicki-Garcia, S. (1969). Cell wall differentiation in the Phycomycetes. *Phytopathology* **59**, 1065–1071.

Bartnicki-Garcia, S. (1973). Fundamental aspects of hyphal morphogenesis. *Symposium Society of General Microbiology* **23**, 245–267.

Bartnicki-Garcia, S. and Lippman, E. (1969). Fungal morphogenesis: cell wall construction in *Mucor rouxii*. *Science* **165**, 302–304.

Bartnicki-Garcia, S., Nelson, N. and Cota-Robles, E. (1968). A novel apical corpuscle in hyphae of *Mucor rouxii*. *Journal of Bacteriology* **95**, 2399–2402.

Bracker, C. E. (1971). Cytoplasmic vesicles in germinating spores of *Gilbertella persicaria*. *Protoplasma* **72**, 381–397.

Fletcher, J. and Morton, A. G. (1970). Physiology of germination of *Penicillium griseofulvum* conidia. *Transactions of the British Mycological Society* **54**, 65–81.

Florance, E. R., Denison, W. C. and Allen, Jr. T. C. (1972). Ultrastructure of dormant and germinating conidia of *Aspergillus nidulans*. *Mycologia* **69**, 115–123.

Grove, S. N. (1972). Apical vesicles in germinating conidia of *Aspergillus parasiticus*. *Mycologia* **64**, 638–641.

Grove, S. N., Bracker, C. E. and Morre, D. J. (1970). An ultrastructural basis for hyphal tip growth in *Pythium ultimum*. *American Journal of Botany*, **57**, 245–266.

Hawker, L. E. (1966). Germination, morphological and anatomical changes. *In* "The Fungus Spore", 151–161. Ed. Madelin, M. F. Butterworth Science Publications, London.

Hawker, L. E. and Abbott, P. M. (1963). An electron-microscope study of germination of conidia of *Botrytis cinerea*. *Journal of General Microbiology* **33**, 43–46.

Hemmes, D. E. and Hohl, H. R. (1971). Ultrastructural aspects of encystation and cyst-germination in *Phytophthora parasitica*. *Journal of Cell Science* **9**, 175–191.

Lessie, P. E. and Lovett, J. S. (1968). Ultrastructural changes during sporangium formation and zoospore differentiation in *Blastocladiella emersonii*. *American Journal of Botany* **55**, 220–236.

Manners, J. G. (1966). Assessment of germination. *In* "The Fungus Spore", pp. 165–173. Ed. Madelin, M. F. Butterworth Science Publications, London.

Reichle, R. E. and Fuller, M. S. (1967). The fine structure of *Blastocladiella emersonii* zoospores. *American Journal of Botany* **54**, 81–92.

Richmond, D. V. and Pringie, R. J. (1971). Fine structure of germinating *Botrytis fabae* Sardina conidia. *Annals of Botany* **35**, 493–500.

Soll, D. R., Bromberg, R. and Sonneborn, D. R. (1969). Zoospore germination in the water mold, *Blastocladiella emersonii*. 1. Measurement of germination and sequence of subcellular morphological changes. *Developmental Biology* **20**, 183–217.

Tokunaga, J. and Bartnicki-Garcia, S. (1971). Cyst wall formation and endogenous carbohydrate utilization during synchronous encystment of *Phytophthora palmivora* zoospores. *Archiv für Mikrobiologie* **79**, 283–292.

Macromolecular Synthesis during Germination

Brambl, R. M. and Van Etten, J. L. (1970). Protein synthesis during fungal spore germination. V. Evidence that the ungerminated conidiospores of *Botryodiplodia theobromae* contain messenger ribonucleic acid. *Archives of Biochemistry and Biophysics* **137**, 442–452.

Gong, G. C. S. and Van Etten, J. L. (1972). Changes in soluble ribonucleic acid polymerases associated with the germination of *Rhizopus stolonifer* spores. *Biochimica et Biophysica Acta* **272**, 44–52.

Cottlier, D. (1966). Biosynthetic processes in germinating spores. *In* "The Fungus Spore", pp. 217–233. Ed. Madelin, M. F. Butterworth Science Publications, London.

Hollomon, D. W. (1969). Biochemistry of germination in *Peronospora tabacina* (Adam) conidia: evidence for the existence of stable messenger RNA. *Journal of General Microbiology* **55**, 267–274.

Horikoshi, K. and Ikeda, Y. (1969). Studies on the conidia of *Aspergillus aryzae*. IX. Protein synthesizing activity of dormant conidia. *Biochimica et Biophysica acta* **190**, 187–192.

Lovett, J. S. (1963). Chemical and physical isolation of 'nuclear caps' isolated from *Blastocladiella* zoospores. *Journal of Bacteriology* **85**, 1235–1246.

Lovett, J. S. (1968). Reactivation of ribonucleic acid and protein synthesis during germination of *Blastocladiella* zoospores and the role of the ribosomal nuclear cap. *Journal of Bacteriology* **96**, 962–969.

Myers, R. B. and Cantino, E. C. (1971). DNA profile of the spore of *Blastocladiella emersonii*: evidence for gamma-particle DNA. *Archiv für Mikrobiologie* **78,** 252–267.

Ramakrishnan, L. and Staples, R. C. (1970). Evidence for a template RNA in resting uredospores of the bean rust fungus. *Contributions of the Boyce Thompson Institute* **24,** 197–202.

Schmoyer, I. R. and Lovett, J. S. (1969). Regulation of protein synthesis in zoospores of *Blastocladiella*. *Journal of Bacteriology* **100,** 854–864.

Soll, D. R. and Sonneborn, D. R. (1971). Zoospore germination in *Blastocladiella emersonii*: cell differentiation without protein synthesis? *Proceedings National Academy of Sciences U.S.A.* **68,** 459–643.

Soll, D. R. and Sonneborn, D. R. (1971). Zoospore germination in *Blastocladiella emersonii*. III. Structural changes in relation to protein and RNA changes. *Journal of Cell Science* **9,** 679–699.

Soll, D. R. and Sonneborn, D. R. (1972) Zoospore germination in *Blastocladiella emersonii*. IV. Ion control over cell differentiation. *Journal of Cell Science* **10,** 315–333.

Staples, R. C., Bedigian, D. and Williams, P. H. (1963). Evidence for polysomes in extracts of bean rust uredospores. *Phytopathology* **58,** 151–154.

Truesdell, L. C. and Cantino, E. C. (1970). Decay of gamma particles in germinating zoospores of *Blastocladiella emersonii*. *Archiv für Mikrobiologie* **70,** 378–392.

Van Etten, J. L. (1969). Protein synthesis during fungal spore germination. *Phytopathology* **59,** 1060–1064.

Van Etten, J. L. and Brambl, R. M. (1968). Protein synthesis during fungal spore germination. II. Amino-acyl soluble ribonucleic acid synthetase activities during germination of *Botryodiplodia theobromae* spores. *Journal of Bacteriology* **96,** 1042–1048.

Wong, R. S. L., Scarborough, G. A. and Borek, E. (1971). Transfer ribonucleic acid methylases during the germination of *Neurospora crassa*. *Journal of Bacteriology* **108,** 446–450.

Yanagita, T. (1957). Biochemical aspects of germination of conidiospores of *Aspergillus niger*. *Archiv für Mikrobiologie* **26,** 329–344.

5 The Vegetative State

5.1 Introduction

The fungi are one of the most plastic forms of life and exist in a myriad of shapes and sizes, ranging from the minute yeasts and water moulds to the macroscopic mushrooms. The essential vegetative structure of the fungi is the hyphal filament or its equivalent in the unicellular forms, the yeast cell or the chytrid thallus. Hyphae present a large surface area through which substances can be interchanged with the environment. Materials essential for the biosynthetic processes of growth and development are taken up from the environment and waste products excreted back into the environment. Ecologically, fungi are a highly successful form of life and are able to grow in environments hostile to most other forms of life.

5.2. Hyphal Differentiation

The cellular form of fungi is largely determined by the presence of the rigid outer cell wall. A complete understanding of morphogenesis in the fungi requires an understanding of the structure of this wall and the apex which generates it. Furthermore, since a chemically complete cell wall is seemingly essential for normal morphogenesis it is possible to predict that morphogenetic development may depend to some extent on specific variations in the relative proportions of, and/or, the interactions among structural components of the cell wall (Bartnicki-Garcia, 1968a).

The Cell Wall

There are only a few examples in the fungi such as the amoeboidal or flagellar stages of some Myxomycetes and Phycomycetes in which part of a growth cycle exists without a rigid outer wall. Within the life cycle of most fungi there are many morphological growth forms, feeding mycelium, colonizing mycelium, asexual and sexual reproductive forms etc., all of which are achieved by cellular control of wall construction. Thus morphological development within the fungi may be reduced to a question of cell wall morphogenesis.

The fungal cell wall should not be considered as an inert outer skeleton but rather as a complex dynamic zone, the site of diverse enzymatic activity and intimately involved in and responsible for cellular morphogenesis. The recent advent of highly refined techniques for preparing pure cell walls has provided valuable information on the nature of the macromolecular components of the wall fabric. In addition to the structural macromolecules of chitin or cellulose, cell walls contain many other polysaccharides as well as specific amounts of proteins and lipids. Electron microscopy has revealed the spatial arrangements of some of the macromolecular aggregates and how these can vary

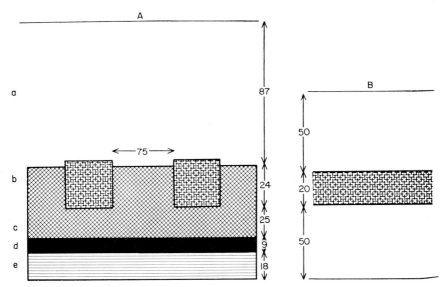

Fig. 5.1. [A] Reconstruction of section through the wall of a hypha from a 5-day culture of *Neurospora crassa* based on enzyme dissection experiments. The numbers represent the mean thickness of the layers in nm. (*a*) Outermost layer of amorphous glucan containing β-1,3 and β-1,6 linkages. (*b*) Sections through coarse strands of (?) glycoprotein reticulum. (*c*) Easily removable protein in which reticulum is embedded; there is an increasing concentration of protein from the outer part of this region inwards. (*d*) A discrete layer of protein. (*e*) Innermost layer of chitin microfibrils possibly intermixed with protein.

[B] Layers visible in a section of an untreated wall from a 5-day culture fixed with glutaraldehyde/O_8O_4; the middle layer is more electron-opaque than the other two. From Hunsley and Burnett (1970).

between genera (Hunsley and Burnett, 1970). Using shadow-cast preparations they have been able to infer the co-axial distribution of certain wall polymers in some filamentous fungi by applying certain sequences and combinations of enzymes (laminarinase, pronase, cellulase and chitinase) which degrade the cell wall from the outside inwards. As a result of these studies it is now possible to achieve a clearer understanding of the chemical and structural basis of the fungal cell wall (Fig. 5.1.)

The fungal cell wall is built up of interwoven microfibrils embedded in or cemented by an amorphous matrix material. In filamentous fungi the microfibrillar or skeletal components of the wall are composed of chitin or cellulose whereas in yeast fungi non-cellulosic glucans make up the skeletal part (Bartnicki-Garcia, 1968a). The cementing matrix substances which bind together the different structural components of the wall into the rigid macromolecular complex are primarily poly-saccharides (mannans, galactan, glucan and heteropolysaccharides) together with protein.

Polysaccharides, which make up about 80–90% of the dry matter of fungal cell walls, are composed of amino sugars, hexoses, hexuronic

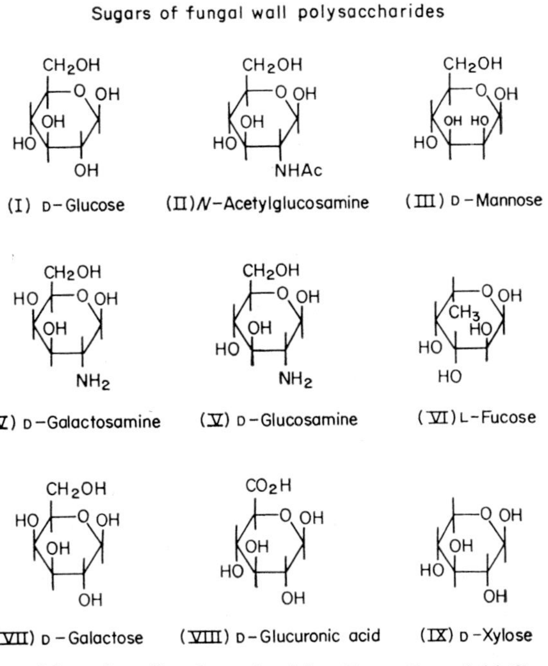

Sugars of fungal wall polysaccharides

Fig. 5.2. Sugars of fungal wall polysaccharides. From Bartnicki-Garcia (1970).

TABLE 5.1. Polysaccharides of fungal walls. From Bartnicki-Garcia (1970)

	Linkages	Monomers
Aminopolysaccharides		
Chitin	β-1,4-	*N*-acetylglucosamine
Chitosan	β-1,4-	D-glucosamine
Neutral polymers		
Cellulose	β-1,4-	D-glucose
β-Glucan	β-1,3; 1,6-	D-glucose
α-Glucan	α-1,3; 1,4-	D-glucose
Glycogen	α-1,4-	D-glucose
Mannan	α-1,2; 1,6; 1,3-	D-mannose
Polyuronides		
Pullularia hetero-	α-1,6; 1,5-	D-glucose,
polysaccharide	β-1,3; 1,6-	D-glucuronic acid,
		D-galactose,
		D-mannose
Mucoran		L-fucose, D-mannose,
		D-galactose,
		D-glucuronic acid

acids, methylpentoses and pentoses (Fig. 5.2. and Table 5.1.). Glucose and *N*-acetylglucosamine are by far the most common building units while D-mannose occurs widely in small amounts in most mycelial fungi and abundantly in yeasts. D-Galactosamine, L-fucose, D-glucosamine, D-galactose, D-glucuronic acid and D-xylose are restricted to certain groups of fungi. Other sugars such as rhamnose, arabinose and ribose occur in such small amounts that it is doubtful whether they are true components of cell walls.

A remarkable feature of the distribution of the polysaccharides in fungal walls is that they appear with unerring regularity in composition and proportion in certain groups of fungi. Bartnicki-Garcia (1968a) has shown that by selecting dual combinations of the polysaccharides which appear to be the main components of fungal cell walls a classification of fungi into a number of cell wall categories can be achieved. Indeed, a close correlation can be observed between cell wall chemistry and the conventional classification based on morphological characteristics (Table 5.2). Most of the septate fungi, i.e. Ascomycetes, Basidiomycetes and Deuteromycetes fall into category V in which chitin and non-cellulosic β-glucans are the major wall components. The yeasts show substantially different wall composition from the other higher

TABLE 5.2. Cell wall composition and taxonomy of fungi*

Cell wall category	Taxonomic group	Representative genera
I. Cellulose-Glycogen	Acrasiales	Polysphondylium, Dictyostelium
II. Cellulose-β-Glucan	Oomycetes	Phytophthora, Pythium, Saprolegnia
III. Cellulose-Chitin	Hyphochytridiomycetes	Rhizidiomyces
IV. Chitin-Chitosan	Zygomycetes	Mucor, Phycomyces, Zygorhynchus
V. Chitin-β-Glucan	Chytridiomycetes	Allomyces, Blastocladiella
	Euascomycetes	Aspergillus, Neurospora, Histoplasma
	Homobasidiomycetes	Schizophyllum, Fomes, Polyporus
VI. Mannan-β-Glucan	Hemiascomycetes	Saccharomyces, Candida
VII. Chitin-Mannan	Heterobasidiomycetes	Sporobolomyces, Rhodotorula
VIII. Galactosamine-Galactose polymers	Trichomycetes	Amoebidium

* For specific examples and references see Bartnicki-Garcia (1968a). Seemingly, the β-glucan is always 1,3- and 1,6-linked. Deuteromycetes are included in their corresponding groups of perfect fungi.

fungi. Within the Phycomycetes there is a wide variety of cell wall composition (Table 5.2).

Thus there exists a remarkable consistency in cell wall composition among related fungi. Polysaccharides involved in wall synthesis do not appear free in the cell but always chemically complexed with proteins and lipids. The formation of cell walls must then be a very complex and elaborate process involving interrelationships between enzymes involved in precursor formation, transport of precursors, and their polymerization into three-dimensional structures with the shape characteristic of the phase of development.

A novel approach to the problem of wall synthesis in fungi has been the studies of wall regeneration by fungal protoplasts. Protoplasts can now be readily obtained from many filamentous fungi and under suitable conditions these protoplasts can revert to the filamentous morphological form from which they were produced (Peberdy, 1972). The process of regeneration is complex but undoubtedly offers exciting prospects for studies on wall synthesis and on the factors controlling the differentiation of cellular form.

The Hyphal Apex

Hyphal tips have structural and cytochemical properties which make them distinct from the rest of the thallus. Brunswik (1924) first observed with haematoxylin-stained preparations of two species of *Coprinus* that the hyphal tips contained a strongly staining area at the growing point. He was unable to show conclusively that it was of "nuclear" composition and eventually considered it to be a special organelle connected in some way with apical growth. This organelle was called the Spitzenkörper and could be demonstrated in living mycelium by using phase-contrast microscopy (Fig. 5.3.). When

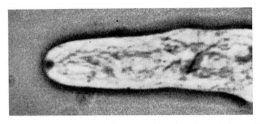

Fig. 5.3. Phase-contrast micrograph of a growing hypha of *Aspergillus niger* showing the dark Spitzenkörper (S) within the light apical area. The Spitzenkörper is not touching the apical wall. From Grove and Bracker (1970).

extension growth stops the Spitzenkörper disappears to reappear again prior to commencement of extension growth. The position of the Spitzenkörper in the apical dome is related to the subsequent direction of extension—an eccentric position preceding a turning of the tip. Spitzenkörpers have now been identified in all septate fungi (Ascomycetes, Deuteromycetes and Basidiomycetes) so far examined (McLure, Park and Robinson, 1968; Grove and Bracker, 1970) though none has ever been observed in the Phycomycetes.

Electron microscopic studies of the Oomycete fungus *Pythium* sp. have shown that a young hypha consists of three zones: an apical zone, a subapical zone and a zone of vacuolation. The apical zone is characterized by an accumulation of cytoplasmic vesicles often to the exclusion of other organelles and ribosomes (Fig. 5.4). In some instances it is possible to observe vesicle membranes which are continuous with the plasma membrane. The subapical zone is not vacuolated and is characteristically rich in a variety of protoplasmic components: nuclei, mitochondria, ribosomes, endoplasmic reticulum, dictyosomes, vesicles, microbodies and microtubules (Fig. 5.5). Dictyosomes are positioned adjacent to endoplasmic reticulum or nuclear envelope and vesicles

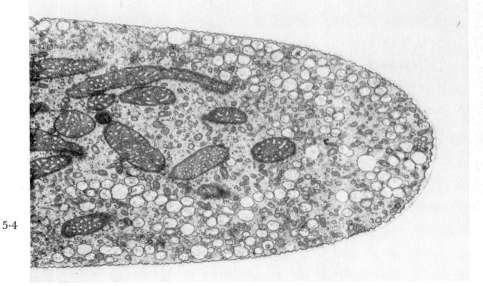

5·4

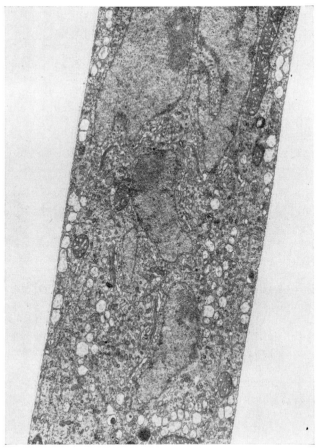

5·5

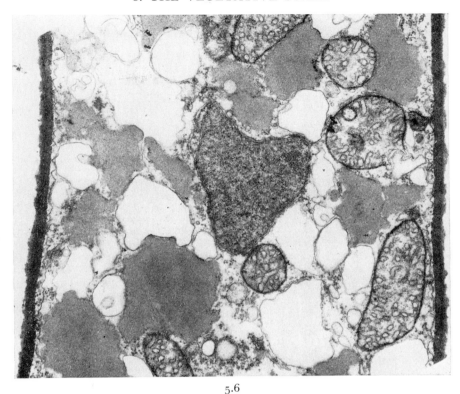

5.6

Figs. 5.4–5.6. Electron micrographs of longitudinal sections through hyphae of *Pythium ultimum*. Fig. 5.4. The apical zone is rich in vesicles, and the mitochondria are clustered in the posterior portion of this zone. Note the scarcity of ribosomes and ER compared with Fig. 5.5. Sub-apical zone showing the complement of cell components and their distribution. A few lipid bodies are found in this zone. Fig. 5.6. Vacuolated zone of a young hypha. From Grove, Bracker and Morré (1970).

occur at the peripheries of the dictyosomes. Some distance from the hyphal apex the subapical zone merges into the zone of vacuolation (Fig. 5.6). The degree of vacuolation increases with distance from the apex and is paralleled by an increase in lipid content. The origin of the vesicles and their movement to the hyphal apex is shown diagrammatically in Fig. 5.7. It has been considered that the apical vesicles are involved in wall synthesis, forming in the subapical regions and moving to the apex where they secrete their contents or fuse with the plasma membrane. Such apical vesicles could also contain exoenzymes which could be extruded at the hyphal tip.

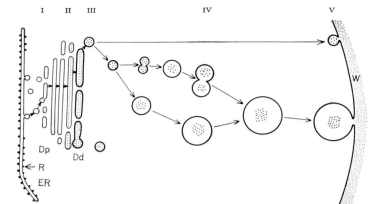

Fig. 5.7. Diagrammatic interpretation of the sequence leading to expansion of a hypha at the apex. (I) Material is transferred from ER to dictyosome by blebbing of ER and refusion of vesicles to form a cisterna at the proximal pole of the dictyosome (Dp). (II) Cisternal contents and membranes are transformed as the cisterna is displaced to the distal pole (Dd) by the continued formation of new cisternae. (III). Cisternae vesiculate to form secretory vesicles as they approach and reach the distal pole. (IV). Secretory vesicles migrate to the hyphal apex. Some may increase in size or fuse with other vesicles to form large secretory vesicles, while others are carried directly to the cell surface. (V). Vesicles accumulate in the apex and fuse with the plasma membrane, liberating their contents into the wall region (W). From Grove, Bracker and Morré (1970).

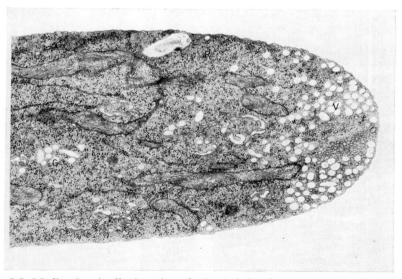

Fig. 5.8. Median longitudinal section of a hyphal tip of *Aspergillus niger*. Amidst the cluster of apical vesicles (V) there is a zone consisting of an aggregate of smaller vesicles and clusters of ribosomes. The larger vesicles do not intrude into this area. Note the abrupt decline in the concentration of ribosomes (R) in the apical zone as compared to the subapical portion in which apical vesicles are not concentrated. Smooth-surfaced Golgi cisternae (G) and a few vesicles occur immediately basipetal to the apical zone. ×20,000. From Grove and Bracker (1970).

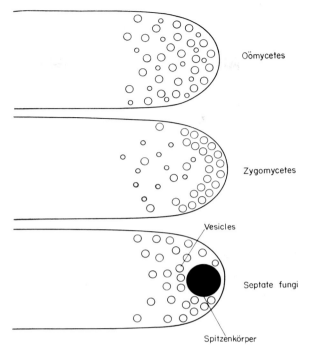

Fig. 5.9. Diagrammatic comparisons of the principal forms of apical organization in hyphae, based on representatives from major tazonomic groups. From Grove and Bracker (1970).

Grove and Bracker (1970) have conclusively demonstrated that the Spitzenkörper is the light microscope equivalent of a special zone of corpuscular material at the tip. The Spitzenkörper is now identified with the denser central part of the apical cluster of vesicles which are characterized by numerous microsomal tubules (Fig. 5.8). The three distinct patterns of vesicle aggregation in the hyphal tips of fungi are shown in Fig. 5.9. The interrelationship of vesicle formation, polymer synthesis, transport and insertion are as yet not well understood.

Hyphal Extension

That hyphae extend by synthesizing new wall at the apex has been shown by microscopy (Robertson, 1965), autoradiography (Bartnicki-Garcia and Lippman, 1969; Gooday, 1971; Katz and Rosenberger, 1971), and by fluorescent antibodies specific for wall components (Marchant and Smith 1968). While the hyphae increase in length,

localized sites of wall synthesis form at subapical positions giving rise
to lateral branches where the apical dominance of wall synthesis again
prevails. Subapical sites can also function for a limited period of time
and form the simple septa found in the Ascomycetes or the more complex parenthesome of the Basidiomycetes (Moore, 1965) Although wall
extension is mainly restricted to the apex of all growing hyphae, wall
thickening may occur for a short distance behind the apex and
occasionally at distal parts, e.g. chlamydospore formation.

Several studies have been made on the incorporation of radioactive
glucose and/or N-acetylglucosamine into the wall fabric of growing
fungal hyphae. Autoradiographic studies with glucose or N-acetyl-
glucosamine as substrate have shown that most of the incorporation
occurs at the extreme tip of the hypha (Fig. 5.10).

The maximum rate of incorporation of radioactive wall precursors
per unit area occurs at or within 1 μ of the apex and decreases sharply
over a short distance corresponding approximately to the length of the

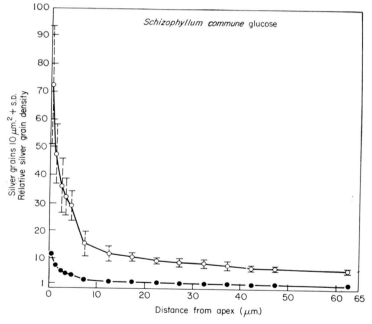

Fig. 5.10. Incorporation of radioactive glucose into the wall fabric of growing fungal
hyphae. Counts of silver grains from autoradiographs of *Schizophyllum commune*
incubated for 10 min. in [5 – ^{3}H] D – glucose. $\overline{\underline{Q}}$, Silver grains/10μm^2 ± standard
deviation (10 tips counted). ●, Silver grain count relative to that 50 to 75 μm behind
the apex. From Gooday (1971).

apical dome. There is a residual gradient of wall deposition in the tubular portion of the hypha and this probably accounts for the increase in growth, thickness or microfibril diameter.

In these autoradiographic experiments mycelium was exposed to tritiated precursors for short periods of time, subsequently killed, and treated in such a way as to remove all cytoplasmic radioactivity without destroying the original shape of the cell. The resulting cell ghosts were then stained, fixed on a microscope slide, coated with nucleare mulsion and processed for autoradiography. When viewed under a light microscope the silver grains correspond almost entirely to glucosamine or acetylglucosamine molecules incorporated into the cell-wall polymers.

Recently, Katz and Rosenberger (1971) have shown that inhibition of cellular protein synthesis by cycloheximide treatment or by an osmotic shock rapidly and drastically altered the normal pattern of apical incorporation of N-acetylglucosamine in *Aspergillus nidulans* to one of insertion along the length of the hypha. Such treatments also produced hyphae with abnormally large numbers of branches and septa. This subapical incorporation of N-acetylglucosamine induced by cycloheximide was readily reversed by the removal of the inhibitor. Since cycloheximide is a potent inhibitor of enzyme formation newly synthesized enzymes could not be responsible for the change in the pattern of incorporation. Thus it must be considered that pre-existing enzymes located along the length of the wall are being activated by cycloheximide. The nature of the postulated activation remains completely unidentified and could well be one of several possibilities such as an increased supply of substrate, exposure of a priming molecule, or the removal of an inhibitor.

The studies with the radioactive labelled wall precursors have conclusively shown the site of final deposition but not the site of synthesis. Electron microscope studies would suggest that the vesicles do not transport wall fibrils and it is now almost certain that the microfibrillar skeleton of the wall is synthesized *in situ* either on the outer surface of the plasmalemma or within the wall fabric. The amorphous matrix material is probably synthesized in the cytoplasm behind the apex, transported there in the vesicles, and discharged and directly anchored to the fibrillar material.

That the fungal wall is both chemically and anatomically an extremely complex multi-component system has already been indicated and it could well be that the different components have different cellular sites of synthesis. Hunsley and Burnett (1970) have shown that the hyphal apices of *Phytophthora*, *Neurospora* and *Schizophyllum* contain randomly orientated microfibrils in their walls. It has been considered by Gooday (1971) that the main incorporation of radioactive glucose or *N*-acetylglucosamine by these fungi may represent microfibril formation. Subapical incorporation on the other hand may imply intussusception since Hunsley and Burnett (1968) showed that the dimensions of the microfibril elements in the walls of these fungi increased away from the tip.

Lytic Enzymes and Apical Growth

It has long been considered that lytic enzymes may be involved in the overall process of apical growth. However, their implication in hyphal growth has not yet been conclusively demonstrated although much circumstantial evidence is available, viz: (*a*) the widespread ability of hyphae to fuse or anastomose by self-dissolution and the phenomenon of clamp-connection formation in the Basidiomycetes: (*b*) positive correlation between the level of cell-wall lytic enzymes (proteases, glucanases etc.) and apical growth; (*c*) germination processes of spores involving wall softening; (*d*) autolytic breakdown so readily seen in many of the sporophores of the Basidiomycetes.

If lytic enzymes do perform an essential function in apical growth it is not unlikely that the enzymes responsible may also be discharged at the apex from the vesicles in a manner similar to the wall synthesizing enzymes. Whether they come in the same or different vesicles will have to be determined.

Assuming that lytic enzymes are involved in apical growth the question now arises as to what controls the balance between wall synthesis and wall lysis. Wall synthesis in the absence of lysis could cause excessive wall thickening and possible arrestment of growth whereas lysis in the absence of synthesis would result in bursting the hyphal tips. It is of interest to note that the addition of polyoxin D (a potent inhibitor of chitin synthetase) and chitinase (an enzyme which will degrade chitin) to fungal cultures have the same effect of promoting the excessive bursting of hyphal tips.

The harmonious balance between the factors responsible for wall synthesis and wall lysis must undoubtedly be involved not only in hyphal tip growth but also in determining the ultimate shape of the fungal cell. Thus a better understanding of the interrelationships between these factors must undoubtedly lead to a fuller understanding of hyphal and indeed fungal morphogenesis.

Numerous hypotheses have been put forward to explain the mechanism of apical growth in fungi. Recently, Bartnicki-Garcia (1973) has presented a unitary model of cell-wall growth which considers that wall growth results from the cumulative action of minute hypothetical units of wall growth. The essence of this hypothesis is that all examples of wall growth can be more readily explained on the basis of the distribution of these units. Fig. 5.11 shows the proposed mechanism involved in the increase in size of such a hypothetical minimum unit. For simplicity the model considers that the fungal wall is composed of two major components: an amorphous substance and a microfibrillar skeleton.

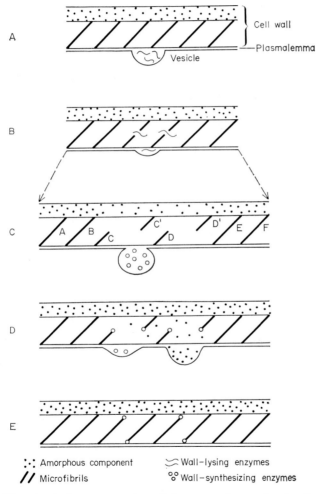

Fig. 5.11. Hypothetical representation of the events in a unit of cell-wall growth. For detailed explanation see text. From Bartnicki-Garcia (1973).

Five major events are considered to occur to allow increase in wall growth:

A. The process is initiated by the secretion of lytic enzymes from a vesicle into the wall fabric.

B. The lytic enzymes attack the microfibrillar skeleton by splitting either inter- or intramolecular bonds and so dissociating the microfibrils.

C. The weakened microfibrillar complex is not able to withstand the high turgor pressure of the cell and becomes stretched and less integrated, thus allowing an increase in surface area to occur.

D. The dissociated microfibrils are rebuilt by synthetase enzymes situated in the wall itself or on the outer surface of new plasma-lemma. The synthetase enzymes are initially secreted into the wall fabric from vesicles, and the soluble precursors are assumed to be transported across the plasmalemma. Vesicles containing the amorphous wall material in a largely or entirely preformed state deposit their contents against the wall and this material is then forced into the microfibrillar network by turgor pressure.

E. At the completion of this enzymic interplay the cell-wall unit has expanded one unit area without losing its overall properties in particular, the coaxial arrangement of wall polymers.

By means of this model apical growth would be the result of the concentration of the growth units around a point of the cell surface that was to become the apical pole of a hyphal tube. How such polarity is maintained to allow tubular growth cannot yet be explained. The ultimate shape and diameter of a hypha would be determined by the spatial distribution of the postulated growth units and by the relative ratios of biosynthetic and lytic functions in these units. Hyphal morphology could then be strongly influenced by a broad spectrum of environmental conditions as well as by mutations at particular loci. Much experimental work must yet be done to prove fully this brilliant hypothesis, but it clearly offers an experimental system which may lead to a better understanding of many of the current problems in fungal morphogenesis.

5.3 Vegetative Morphogenesis

Colony Formation

At the hyphal apex extension in length is potentially unlimited while increase in width is strictly limited. At varying distances behind the apex branch hyphal apices are initiated and in septate fungi cross-walls may be formed. One of the most interesting features of hyphal branching is that the branch from any given hypha does not necessarily become a cell with the same features as the parent cell. Growth and differentiation of individual hyphae take place with varying degrees of co-ordination with adjacent hyphae, and in some forms more or less organized multihyphal aggregates may develop. The individual hyphae of many fungi grow at a constant rate, i.e. their length in-creases linearly with time.

On static or liquid surface culture most filamentous fungi produce a heterogenous growth form which includes aerial, surface and sub-merged hyphae each with its particular physiological condition. Studies of surface cultures of *Aspergillus niger* have shown that cytological characteristics and physiological activities of the hyphae vary from the periphery to the inside portions. From the differential localization of enzymes, wall thickness and intracellular structures of the hyphae it appears that hyphal cells show a characteristic differentiation depend-ing on their localization in the culture. Autoradiographic studies of nucleic acid and protein metabolism in individual hyphae of several fungi clearly demonstrates the heterogeneous nature of the hyphal filament (Nishi *et al.*, 1968). This differentiation of cellular activities along the hyphal filament is probably an inevitable consequence of the apical growth process.

The control of the growth rate of fungal colonies on surface culture has been studied by Trinci (1971) who has presented a hypothesis which relates the linear growth observed in surface colonies to the exponential growth which can be obtained in liquid culture (see also Chapter 8 and Righelato, 1974). The hypothesis assumes that after an initial phase of exponential growth, the growth at the centre of the colony becomes restricted by the slow rate of diffusion of nutrients (Fig. 5.12). When this occurs the rate of growth of the colony becomes dependent upon the specific growth rate of the peripheral hyphae and the width of the peripheral zone in which growth is not limited by the rate of diffusion of nutrients. If the growth of the colony can be attributed to exponential growth of the mycelium in the outer annulus then the radial growth rate can be related to the specific growth rate by the following equations:

$$\frac{dr}{dt} = \mu a$$

$$r_t = a\mu t + r_o$$

where μ = specific growth rate
a = width of peripheral growing zone
r_o and r_t = radius of colony at time zero and time t.

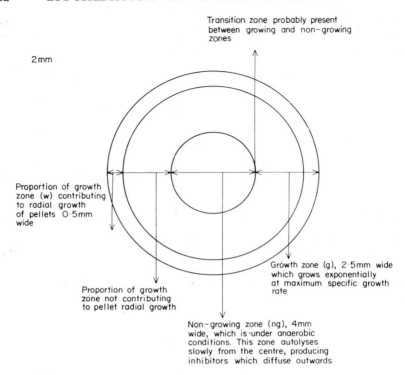

Fig. 5.12. Proposed zones of a hypothetical pellet of *Aspergillus nidulans* grown at 30°C and 9 mm in diameter. From Trinci (1970).

From these equations it can be seen that to calculate the specific growth rate of a fungal colony it is necessary to know both the linear rate of colony growth dr/dt and the width of the peripheral growing zone (a). This value has been estimated by cutting a fungal colony along a chord and studying the rate of growth of the mycelium within this chord. When the distance between the incision and the edge of the colony is less than a the effect of the incision will be to reduce the growth rate. The value of a can be measured as the shortest distance between the incision and the edge of the colony at which the incision had no effect on growth rate. Using values of a estimated in this manner, values of μ have been calculated which correspond closely with those obtained in liquid culture. The value of a and hence the radial growth rate have been found to be lower in strains of *Aspergillus nidulans* with a high frequency of branching whereas the value of μ was unaffected.

Because of the heterogeneous nature of mycelial growth in surface culture most physiological studies have been made in submerged culture which provides a 3-dimensional relatively uniform environment for growth. The growth kinetics of filamentous fungi in liquid culture are complex and vary with the fermentation technique being used, whether continuous or batch, and with the growth form of the organism, i.e. filamentous or pellet. This aspect will be fully developed in Chapter 8.

Special Multihyphal Structures

In many filamentous fungi multihyphal vegetative structures are formed that appear to involve some degree of co-ordination between the component hyphae that is greater than that found in a vegetative colony. Such structures can arise in several ways: by differentiation with neighbouring hyphae, by interhyphal contact to form a solid tissue, by formation of a structure with a definite boundary, or by a period of existence as a more or less discrete structure (Butler, 1966).

Fungal strands may be considered as linear hyphal aggregates which are capable of extending in one direction. These strands are widely found in Basidiomycete fungi and arise from vegetative mycelium, and their normal function is to serve as links between nutrient supply and fruit-bodies.

The strands are predominantly of two types (*a*) mycelial strands or (*b*) rhizomorphs. The mycelial strand is formed by the gradual build-up around a predominantly mycelial frame-work, whereas the rhizomorph is formed by the predominantly apical extension of associated hyphae. In each type of strand there may be varying degrees of hyphal differentiation in the form of wall thickening and subsequent reduction of cell lumen or by the increase in width of intercolony hyphal segments.

Little is known about the environmental conditions that initiate the formation of these ecologically important vegetative structures and almost nothing is known about their internal biochemistry. This is indeed surprising when it is considered that these vegetative strands play an essential part in many diseases of trees and also in the depredations of the dry-rot fungus *Merulius lacrymans*. A better understanding of the initiation of these strands must surely be advantageous in developing methods for their control.

Biochemical Genetics of Vegetative Development

Morphological mutants which grow as fast or faster than the wild type are frequently recognized as sectors in colonies growing on agar plates. On the other hand by this technique slow growing mutants are rapidly overgrown by the wild type and are not readily recognizable. They can however be isolated by growing colonies from single spores which have been exposed to a mutagenic treatment. Just as the isolation of strains which are auxotrophic for a particular metabolite has facilitated the discovery of many biosynthetic pathways, the isolation of morphological mutants should provide a valuable technique for studying morphogenesis.

Morphological mutants have been isolated in many species. However, the most detailed studies have been carried out using *Neurospora crassa* in which mutations at eighty different loci on seven linkage groups have been recognized. One of these, the *colonial* mutant (*col* 2) has a compact growth form and has a different cell wall composition from the wild type (Mehadewan and Tatum, 1965). That this strain is also prototrophic and that the addition of metabolites to the medium has no effect on its growth form indicates that the reaction affected is directly concerned with morphogenesis. Further studies have indicated that the glucose-6-phosphate dehydrogenase in this strain has a decreased thermal stability and a reduced affinity for its substrates- glucose-6-phosphate and NADP. This enzyme is also altered in two other morphological mutants *balloon* and *frost*. In *col* 2, *balloon* and another mutant *col* 3 which has an altered 6-phosphogluconic acid dehydrogenase, it has been demonstrated that the levels of NADH and NADPH are reduced. Glucose-6-phosphate dehydrogenase is the first enzyme in the pentose phosphate pathway which is essential for the generation of high levels of NADPH, so a mutation affecting this enzyme or 6-phosphogluconic acid dehydrogenase, which catalyses the other NADPH generating reaction, would be expected to depress NADPH levels. Since NADPH is involved in many biosynthetic reactions mutations affecting its synthesis will be pleiotropic and the specific reaction causing an observed morphogenetic effect will not be readily identified. A temperature sensitive mutant *col* 2_{b_2} has been isolated which grows normally with normal NADH and NADPH levels at 23°C but exhibits the colonial morphology with reduced NADH and NADPH levels at 34°C (Brody, 1970).

The levels of NADH and NADPH are not affected by a mutation *ragged* (*rg*) which is known to have a defective phosphoglucomutase. In *Neurospora sitophila* two genes *rg* 1 and *rg* 2 have been shown to control two isoenzymes of phosphoglucomutase. *In vivo* studies indicate that the enzyme can exist as two separate molecules or a single complex molecule which it is thought may be the active form *in vivo*. If this is correct then mutations at either *rg* locus could inactivate the enzyme (Misra and Tatum, 1970).

Using acrylamide gel electrophoresis specific changes in the soluble protein fraction have been correlated with and shown to be controlled by a series of morphological mutants at the "*peak*" locus. These changes appear to be related to the change in morphology since the addition of specific nutrients to the medium can stimulate the development of the mutant morphology in the wild type strains. In this case, the soluble protein composition of the mycelium is also altered (Barker *et al.*, 1969).

A fascinating aspect of fungal morphogenesis which deserves more attention from biochemists and geneticists is the frequency of morphological mutants which appear to be inherited cytoplasmically. In *Aspergillus glaucus* spore germination, growth rate, pigmentation and perithecial density have been reported to be under cytoplasmic control. In *Podospora anserina* senescence of the mycelium and barrage formation during incompatible matings also appear to be controlled cytoplasmically (Esser and Kuenen, 1967).

Cytoplasmic inheritance has been well established in the fungi as a result of studies on the mitochondrial mutants *poky* and *petite* and more recently the killer phenomenon in yeast. However, whereas biochemical evidence is now available for the existence of DNA in mitochondria and a double stranded RNA molecule which is associated with the killer phenomenon in yeast (Berry and Bevan, 1972) the existence of specific, potentially self-replicating molecules has not been demonstrated in any of the examples of cytoplasmic morphological mutants. Hypotheses have been presented which account for cytoplasmic inheritance on the basis of stable regulator proteins (Esser and Kuenen, 1967). Further biochemical studies are required to distinguish between these hypotheses and the more conventional view that cytoplasmic inheritance must be attributable to extrachromosomal self-replicating molecules.

5.4   Yeast/Mould Dimorphism

The ability of some fungi to exhibit a phenotypic duality of cell form has provided a valuable approach to a fuller understanding of the biochemical basis of vegetative differentiation. Fungi that show dimorphism are able to exist as filamentous mycelia (M form) or as spherical yeast-like cells (Y form) which can reproduce by budding (Fig. 5.13). Control of dimorphism may be simple or complex involving in some cases only one environmental factor while in others several interrelated environmental factors would appear to be required (Romano, 1966).

Dimorphism was first observed in deep-seated systemic mycotic infections of man and animals and as such was defined as the condition

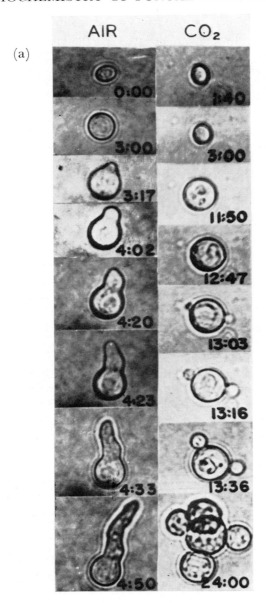

Figs. 5.13 (*a–c*). Dimorphism in *Mucor rouxii*. (*a*). Germination of sporesunder air or under CO_2; incubation time indicated in h; min; × 900. (*b*). Filamentous form of growth under air. × 125. (*c*). Yeast-like form of growth under CO_2. × 625. From Bartnicki-Garcia and Nickerson (1962).

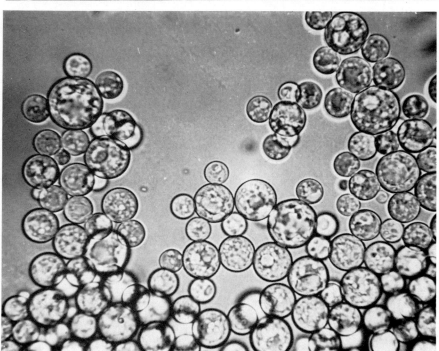

Figs. 5.13 (*b* and *c*)

in which there is a yeast-like parasitic phase and a mycelial saprophytic phase. However, dimorphic forms occur in several non-pathogenic fungi and it would be correct to consider dimorphism as an environmentally controlled reversible interconversion of yeast (Y) and mycelial form (M). As such, fungal dimorphism has been considered to represent a plausible example of primitive morphogenesis (Haidle and Stork, 1966.)

Since so many pathogenic fungi are dimorphic most of the early studies on this phenomenon were medically orientated. Although these worthwhile studies continue there has in recent years been intensive examination of dimorphism in non-pathogenic species, in particular *Mucor*, and as a result there now exists a wealth of scientific knowledge of the biochemical basis of this phenomenon. As yet there is little evidence to suggest that this information could be used in any curative aspect of human and animal mycoses. It is interesting to note that only deep mycotic infections exist in the yeast form. In contrast, dermatophytic fungi which cause superficial infections of the keratinized layers of the skin and fungi which cause plant diseases always exist in the mycelial form. Thus the potential to develop as a yeast may be an important factor in determining the ability of the fungi concerned to establish themselves as internal parasites of warm-blooded animals.

Environmental Control of Dimorphism

Many environmental factors can be involved in determining dimorphism in fungi (Romano, 1966). However, it is now clear that there are three main groups of dimorphic fungi: (*a*) temperature dependent; (*b*) temperature and nutrient dependent; and (*c*) nutrient dependent.

TEMPERATURE DEPENDENT DIMORPHISM

Blastomyces dermatitidis, the causal organism in North American blastomycosis, and *Paracoccidioides brasiliensis* each show morphological dimorphism which is totally dependent on temperature, growing in a mycelial (M) form at 25°C and 20°C respectively and a yeast (Y) form at 37°C. Recent studies on the biochemistry of cell-wall formation in *P. brasiliensis* have illustrated interesting differences between the two cell-wall types. In the Y form the main cell-wall polysaccharide is α-glucan whereas in the M form it is β-glucan and galactomannan. Radiochemical studies have shown that the synthesis of α-glucan decreased rapidly after the incubation temperature was changed from 37° to 20°C. Correspondingly, the synthesis of β-glucan was augmented at an early stage of the morphological change. Cell-free extracts from the Y form contained five times more protein disulphide reductase than the M form whereas the M form contained almost eight times more β-glucanase activity than the Y form. The cell wall of the M form contained twelve times more disulphide linkages than the Y form.

An intriguing and highly probable mechanism for the production of the dimorphic forms in *Paracoccidioides brasiliensis* is shown in Fig. 5.14. Ultrastructurally, the cell-wall of the Y form is considered to be composed of an outer layer of α 1, 3-glucan, an inner layer of chitin, and

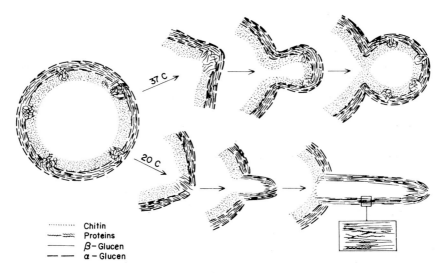

.... Chitin
— ≈≈ Proteins
——— β- Glucen
— — α - Glucen

Fig. 5.14. Proposed hypothesis for the production of yeast-like and mycelian forms of *Paracoccidioides brasiliensis*. For explanation see text. From Kanetsuma *et al.* (1972).

small amounts of β-glucan, proteins and lipids. The small amount of β-glucan is believed not to be evenly dispersed throughout the wall but rather to occur as individual islets. By some as yet unknown mechanism there is a loss of rigidity around the β-glucan and the weakened part is blown out as a bud. It is possible that protein disulphide reductase and β-glucanase may play a role in initiating this change in cell-wall plasticity. At the budding site and also in all parts of the cell-wall of the daughter cell at 37° α-glucan and chitin are synthesized much more actively than β-glucan, resulting in the formation of the spherical Y form.

However, when growth is proceeding at 20° the synthesis of α-glucan decreases at the budding sites and β-glucan fibres extend continuously by end to end linkage leading to the formation of apical and cylindrical growth, typical of the M form. Wall rigidity could well be enhanced by the high amount of disulphide linkages coupled with the low activity of protein disulphide reductase. Chitin may also be mechanically important in the production of the M form since β-glucan cannot be extracted unless there is prior hydrolysis of the chitin. In other words this implies that the fibres of protein and chitin may be interwoven with fibres of β-glucan in the M form cell wall whereas in the Y form cell wall the α-glucan and chitin form two distinctly separate layers.

Temperature and Nutrient Dependent Dimorphism

In this group elevated temperature alone is not sufficient to cause dimorphism; dimorphic expression will only occur when elevated temperature is coupled with the presence of certain nutrients. *In vitro* cultures of *Sporotrichum schenckii* and *Histoplasma farcimonosum* will grow in a yeast-like form at 37° in the presence of 15–20% CO_2. At 37° minus CO_2 or CO_2 at 25° a mycelial growth will result. *Histoplasma capsulatum*, another pathogenic organism, requires a more complex nutrient environment together with elevated temperature to grow in the yeast form. As yet no general scheme has been put forward to explain adequately the mechanism of M → Y conversion in this fungus. It is regrettable that so few biochemical studies have been carried out on this interesting example of dimorphism.

Nutrient Dependent Dimorphism

Dimorphism in this group of fungi is independent of temperature and depends on a variety of environmental factors such as oxygen, carbon dioxide, hexose sugars, heavy metals and other factors that can be present in complex nutrients (for ref. see Romano, 1966). By a suitable selection of these factors it is possible to control proportions of yeast and mycelial forms. Within this group a pathogenic fungus (*Candida albicans*) and a non-pathogenic fungus (*Mucor rouxii*) have been extensively studied.

Dimorphism in Candida Albicans

The normal morphological state of this fungus is the yeast form and only under certain conditions will filamentous growth occur. *In vivo*, growth is normally yeast-like though unlike the other pathogenic fungi that exhibit dimorphism, filamentation can be observed in some infections. When growing saprophytically in the soil or in plant debris the M form is the normal morphological condition.

The studies of Nickerson and others suggest that dimorphism in *Candida albicans* is the result of an interplay between two fundamental cellular processes: cell growth or elongation and cell division or budding. In one set of environmental conditions cell division can be coupled to cell growth resulting in regular budding. When the environmental conditions impair the division process the cells elongate with little or no budding. Sulphydryl compounds have a marked effect on dimorphism increasing budding in the normal form as well as causing budding in a mutant filamentous form. —SH compounds have been

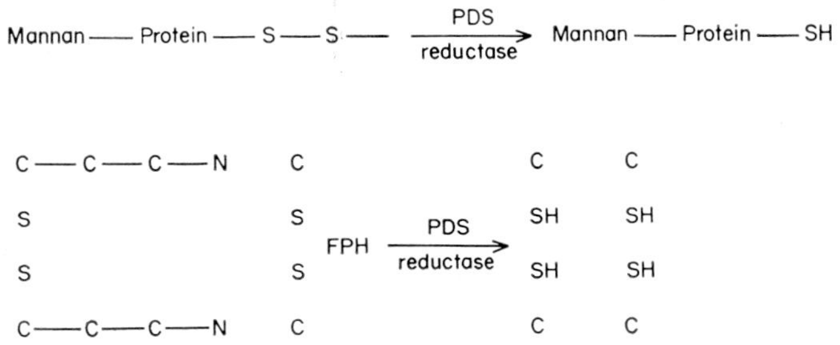

$$\text{Mannan} \longrightarrow \text{Protein} \longrightarrow \text{S} \longrightarrow \text{S} \longrightarrow \xrightarrow[\text{reductase}]{\text{PDS}} \text{Mannan} \longrightarrow \text{Protein} \longrightarrow \text{SH}$$

Fig. 5.15. Dimorphism and protein disulfide (PDS) reductase. Diagram of operation of PDS reductase on disulfide bonds that form covalent links between molecules of the glucomannan-pseudokeratin component of the cell wall. From Falcone and Nickerson (1959).

widely implicated in the division process of many types of cells. Nickerson and Falcone (1956) considered that dimorphism was the result of a change of events that begins with the utilization of metabolically generated hydrogen by a specific cell division enzyme, protein disulphide reductase, for the reduction of disulphide bonds in the mannan-protein complexes of the cell wall. The normal budding strain of *Candida* has a strong protein disulphide reductase activity compared with the slight activity shown by the filamentous mutant form. The reduction within the wall weakens the cell-wall making plastic deformation possible and the subsequent yeast-bud formation can then be considered as a purely physical consequence (Fig. 5.15). It has thus been suggested that in normal dimorphism the M form differs from the Y form in that metabolically generated hydrogen is not coupled to disulphide reduction so the M mycelium is characterized by an excess of reducing power. However, the experimental evidence for these statements is not convincing and a re-investigation of this phenomenon could well prove illuminating.

Dimorphism in Mucor

Certain species of *Mucor*, in particular *M. rouxii*, exist in nature as normal mycelial fungi but do have the potential to develop into budding spherical yeast cells. Bartnicki-Garcia and Nickerson (1962 *a, b*) first demonstrated that a mixture of carbon dioxide and elemental nitrogen was necessary for production of the yeast phase in several species of *Mucor* (Table 5.3). In the absence of carbon dioxide, aerobically or anaerobically, development was typically as a branched coenocytic mycelium. From these studies they concluded that carbon

TABLE 5.3. Morphological response of diverse species and strains of Mucor to different levels of glucose and CO_2 in anaerobic cultivation*

Organisms	100% CO_2		30% CO_2	
	2% Glucose	0·01% Glucose	2% Glucose	0·01% Glucose
Dimorphic strains†				
M. subtilissimus NRRL 1909	Y	Y	Y	Y + M
M. rouxii NRRL 1894	Y	M	Y + M	M
M. racemosus NRRL-1427	Y	M	Y + M	M
M. rouxii IM 80	Y	M	Y + M	M
M. racemosus NRRL 1428	Y	NO	Y + M	M
Nondimorphic strains				
M. hiemalis NRRL 2461	NO	NO	M	M
M. mucedo NRRL 1424	NO	NO	M	M
M. subtilissimus NRRL 1743‡	M	NO	M	M
Actinomucor elegans NRRL 3104	M	NO	M	M
M. pusillus NRRL 1426	NO	NO	NO	NO
M. rammanianus NRRL 1296	NO	NO	NO	NO

* On YPG agar plates incubated for 48 h. Y, yeast development; M, mycelium; NO, no growth.
† Dimorphic strains are listed in decreasing order of yeastlike tendency.
‡ Bartnicki-Garcia (1968). *Journal of Bacteriology* **96,** 1586.

dioxide plays a specific role in the maintenance of yeast growth. Different morphology of growth was correlated with different cell-wall structure, the cell wall of the yeast phase containing six times more mannan than the cell wall of the filamentous phase. Interestingly, mannan has been found to be abundantly present in different species of true yeast and absent in most filamentous fungi (Aronson, 1965).

Bartnicki-Garcia (1963) regarded the crucial difference between M and Y forms to be in the growth polarization. Thus, development of Y represented a selective inhibition or interference with the morphological mechanisms which are indispensable for cylindrical cell formation. Formation of Y was regarded as a consequence of isotropic physical forces.

Using *Mucor rouxii* (NRRL 1894) Haidle and Storck (1966) obtained yeast growth in anaerobic conditions without carbon dioxide and concluded that other nutritional factors were involved in the control of dimorphism. Bartnicki-Garcia (1968 *b*) using *M. rouxii* (IM-80) clearly demonstrated that both hexoses and carbon dioxide are primary determinants of yeast development in *Mucor* spp. and that their dimorphic effects are complementary; at a low pCO_2 a high concentration of hexose is needed for complete yeast-type development and vice versa. If, however, the concentration of hexose in the medium is high enough, carbon dioxide is not required. The effect of glucose concentration could not be attributed to increased production of carbon dioxide since maximal evolution of carbon dioxide is reached with only 0·1% glucose. Hexose also influences the aerobic yeast-like growth of *M. rouxii* (NRRL 1894) and *Candida albicans*.

Phenethyl alcohol, a proven inhibitor of growth in bacterial, fungal and animal cells (Terenzi and Storck, 1969), caused spores of *Mucor rouxii* (NRRL 1894) to form spherical budding cells instead of hyphae provided that the carbohydrate source was a hexose at 2–5%. When the carbohydrate source was xylose, maltose, sucrose or a mixture of amino acids the morphology in the presence of phenethyl alcohol was filamentous. Phenethyl alcohol stimulated carbon dioxide and ethyl alcohol production and inhibited oxidative phosphorylation of extracted mitochondria.

Thus many different environmental factors can promote dimorphic development in *Mucor rouxii* and it is not possible to attribute a causal role to any one factor in particular since the effect of each factor is conditioned by the concentration of the others (Bartnicki-Garcia and McMurrough, 1971). It is probable that these many and varied environmental factors are exerting their morphogenetic influence on the cellular mechanisms that control the polarity of cell wall growth. When wall growth is restricted to one small region (apical growth) tubular hyphal growth will occur whereas when wall growth occurs over the entire periphery of the cell spherical yeast cells will develop.

It is interesting to note that most of the factors that cause yeast type morphology in *Mucor* spp. also favour a fermentative metabolism. Furthermore, there are many other examples where inhibition of aerobic respiration with subsequent enhancement of anaerobic fermentative pathways has led to a restriction of morphological differentiation in filamentous fungi (see also Chapter 6).

Thus these studies do imply that the cellular energy producing systems may be intimately involved in the control of polarity of cell wall synthesis, and perhaps more attention should be given to the relationship between mitochondria and cellular development.

5.5   Vegetative Differentiation in Yeast

Cytological Observations

The group of organisms generally referred to as yeasts do not appear to constitute a natural taxonomic unit but rather a collection of organisms

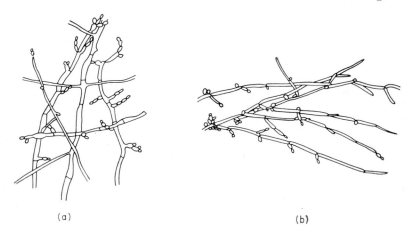

(a)   (b)

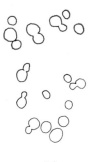

(c)

Fig. 5.16. Mycelial forms in yeast. From Lodder (1970).

(a) True mycelium in *Endomycopsis burtonii*
(b) Pseudomycelium in *Candida tropicalis*
(c) Unicellular growth in *Saccharomyces cerevisiae.*

which have a unicellular growth form for at least one stage in their life cycle, and in general reproduce by budding (Kreger-van Rij, 1969). It appears, however, to be difficult to draw a firm line between yeasts and other filamentous fungi. Under suitable conditions true mycelium may develop in *Endomycopsis*, a dikaryotic mycelium with clamp connections in *Sporidiobolus*, while a pseudomycelium, constructed from branched filaments of single cells, can occur in several genera, e.g. *Candida*, *Hansenula* and *Pichia* (Fig. 5.16). Conversely, a yeast-like growth has also been described in several genera of fungi which are not normally classified as yeasts, e.g. *Mucor*, *Verticillium* and *Ustilago*. Pseudomycelium formation in yeasts is controlled by environmental conditions

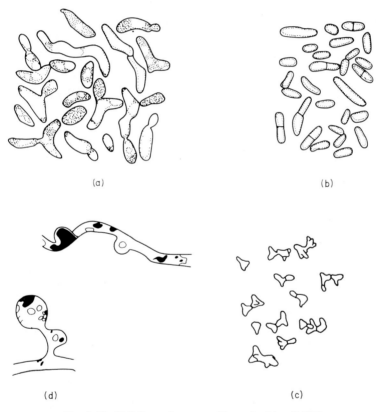

(a) (b)

(d) (c)

Fig. 5.17. Cell forms in yeasts. From Lodder (1970).

(a) *Sporidiobolus holsaticus*
(b) *Schizosaccharomyces pombe*
(c) *Trigonopsis variabilis*
(d) a heterobasidiomycetous yeast.

and can often be recognized macroscopically as a rough, rather than a smooth colony form.

The individual cells of yeast are usually either spherical or ellipsoidal. However, triangular cells occur in *Trigonopsis variabilis* while in the fission yeasts of the genus *Schizosaccharomyces* the cells often have a rectangular outline. The shape of the cells and the detailed morphology of the cells is closely correlated with the method of cell division (Fig. 5.17).

Physiology and Biochemistry of Growth and Reproduction in Yeast

CELL-WALL COMPOSITION

The cell-wall in yeast constitutes about 90% of the dry weight of the cell. Since it is the cell-wall which maintains the structural integrity of the cell and determines the size and shape of the cell, it is not surprising that studies on the biochemistry of the cell wall have made the greatest contribution to an understanding of the control of the vegetative development of the yeast cell. The structure of the yeast cell-wall has recently been covered in a comprehensive review by Pfaff (1971). The major components in the cell wall are mannans, glucans, chitin, proteins and lipids. The glucans and mannans constitute the major structural components in the ascomycetous yeasts, whereas in the basidiomycetous yeasts chitin appears to be the major polysaccharide. Few generalizations can be made about the relative abundance of these carbohydrate fractions since major differences occur between different species and different forms of the same species. The fission yeasts are somewhat distinctive in that they contain no chitin and only a low level of mannan.

Although proteins constitute only 10% of the cell-wall their importance is indicated by the fact that the cell-walls of *Saccharomyces cerevisiae* are much more readily hydrolysed if proteolytic enzymes are added to glucan and mannan hydrolysing enzymes in the incubation mixture. The protein in the cell-wall is complexed with the polysaccharide and is characterized by a high level of sulphur-containing amino acids. These are probably in the form of disulphide bonds which are important in maintaining the structural integrity of the cell-wall. The addition of the reducing agent mercaptoethanol also facilitates the lysis of yeast cell-walls by glucanases and mannanases.

Estimates of the amount of lipid in yeast cell-walls vary from 2% to 13% and probably depend on the degree to which the cell-wall is contaminated with the plasmalemma.

CELL-WALL SYNTHESIS

The problem of development requires that we understand not only the biochemical compounds responsible for maintaining the cell structure, but also the biochemical mechanisms by which the shape of the cell is controlled (Bartnicki-Garcia and McMurrough, 1971). Information concerning the site of bud initiation is discussed in Chapter 6. Conflicting evidence has been presented concerning the role of the cell-wall itself in determining the organization of newly synthesized cell-wall material. It is possible, using mixtures of hydrolytic and proteolytic enzymes in carefully controlled conditions, to remove much of the yeast cell wall without killing the cell, subsequently producing protoplasts if all the cell wall material has been removed and spheroplasts if part remains. Under standard conditions protoplasts cannot regenerate a normal cell wall although they are clearly capable of synthesizing cell wall components such as fibrillar mannans, glucans and proteins which are excreted into the medium. Protoplasts appear to lack some component which is essential for the organization of these components into the cell wall. The incorporation of gelatin or agar into the growth medium has been reported to stimulate cell wall formation by yeast protoplasts. Gelatin and agar would be expected to inhibit the dispersion of the precursor molecules excreted by the protoplasts into the medium. In this system the cell-wall formed in the initial stages of regeneration is disorganized and it is only when the first daughter cells are budded off that normal cell-walls are produced. Disagreement exists however as to whether all the cell wall material has been removed from protoplasts which are able to generate a new cell-wall.

The observation that two polymerases which catalyse the formation of 1–3 β- and 1–6 β-linked glucan from UDP glucose are located in the cell wall indicates that the cell-wall does play an important role in its own biosynthesis. The substrate for these enzymes, UDP glucose, has been shown to be excreted through the plasmalemma in protoplasts of *Saccharomyces fragilis*.

CONTROL OF CELLULAR MORPHOLOGY

The cells of several yeast species show a tendency to form elongated cells under different environmental conditions. Temperature, pH, aeration, levels of carbon and nitrogen substrates and the presence of amino acids and fusel oils have all been implicated. Experiments using continuous culture have shown that elongation occurs when *Saccharomyces cerevisiae* is grown under nitrogen-limiting conditions in aerobic culture. Elongated cells produced in this manner reverted to the normal ellipsoidal form if sodium thioglycollate was added to the medium (Brown and Hough, 1966). A requirement for reduced sulphydryl groups for cell division has been demonstrated. The level of protein in the walls of elongated cells is only half that found in ellipsoidal cells.

Cell elongation can be stimulated in *Hansenula schnegii* by tryptophan in conditions in which the addition of other amino acids had no effect. Tryptophan may be acting in a specific manner as a precursor of in-dole acetic acid (IAA). This plant hormone which is known to be in-volved in cell extension in plants is synthesized from tryptophan and is probably produced by fungi (Turner, 1971). IAA itself has been

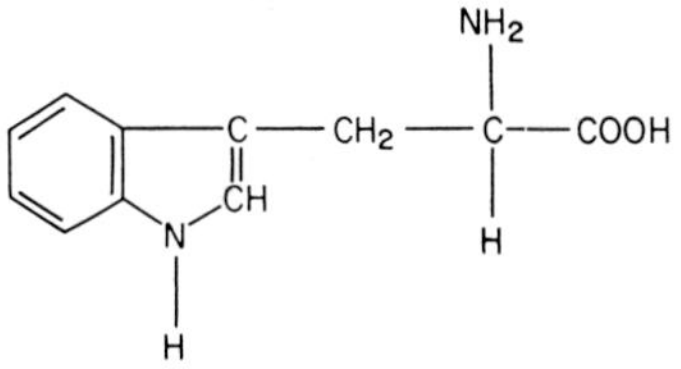

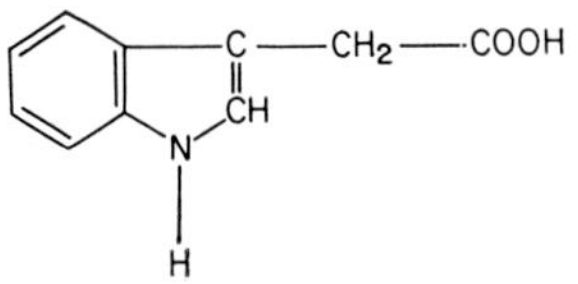

Fig. 5.18. Structure of indole acetic acid and tryptophan.

reported to induce cell elongation in *Saccharomyces cerevisiae* (Shimoda and Yanagishima, 1969; Fig. 5.18).

The production of triangular-shaped cells by *Trigonopsis variabilis* can be stimulated by the incorporation of methionine or choline into the growth medium. Triangular cells produced in this way have a much higher level of phospholipid in the cell-wall fraction (Fig. 5.19) (Senthe shanmuganathan and Nickerson, 1962).

THE CELL CYCLE

"The cell cycle of a growing cell is the period between the formation of a cell by division of its mother cell and the time when the cell itself divides to form two daughter cells" (Mitchison, 1971). Since cells of the same species growing under similar conditions have the same size and chemical composition the level of all the cellular components must

| | Cell form | |
	Ellipsoidal	Triangular
(a) Lipid content of whole organisms mg/100 mg dry wt. Fraction Free lipids:		
(a) Ether soluble	1·05	3·0
(b) CHCl₃ soluble	0·20	0·30
(c) Acetone insoluble fraction of (a)	Nil	Trace
Bound lipids	18·7	40·4
(b) Fractionation of bound lipids of whole organisms Bound lipids (g./100g)		
(a) Acetone soluble	96.8	97.1
(b) Acetone insoluble (phospholipids)	3.2	2·9
(c) CHCl₃ insoluble	1·56	4·74
Nitrogenous constituents (mg./100g.)		
(a) Total nitrogen	780·0	560·0
(b) Ethanolamine	70·3	38·6
(c) Choline	37·0	72·0
(d) Serine	66·0	36·0
(e) Choline from lecithin	8·4	29·0
Other properties		
(a) Saponification number	530	315
(b) Acid number	152	86
(c) Ester number (by difference)	378	229
(d) Iodine value	14·0	4·4
(e) Unsaponifiable matter	25 g./100 g.	20·1 g./100 g.
Phosphorus	453	244
	(mg./100 g.)	(mg./100 g.)
N : P ratio	1·72	2·29
Colour test for sterols in unsaponifiable material	+	+++
(c) Bound lipids from cell wall fraction mg./100 gm. Analysis		
Nitrogen	120	420
Phosphorus	100	200
N : P ratio	1·2	2·1
Iodine no.	0	0

Fig. 5.19. Comparison of lipid composition of ellipsoidal and triangular cells of *Trigonopsis variabilis*. Fran Sentheshanmuganathan and Nickerson (1962).

double during each cell cycle. The period of the cell cycle is thus a natural unit of time during which cell duplication occurs. The relationship between the duration of the cell cycle and the increase in the number of cells in a population has been described mathematically and is of considerable importance in fermentation studies:

$$g = \frac{t_2 - t_1}{3 \cdot 32} (\log x_2 - \log x_1)$$

where g = generation or doubling time.
x_1 = concentration of cells at t_1
x_2 = concentration of cells at t_2

The remainder of this section will be concerned with the subcellular events which occur during the cell cycle and which contribute to an understanding of the control of cell growth and division. There are two types of technique available for the study of the cell cycle: (a) the study of the life cycle of a single cell using microscopic techniques and (b) the use of synchronous cultures in which all the cells of a population are at the same stage of the cell cycle at the same time. The limitations of the first technique for biochemical studies are obvious while great care must be exercised when synchronous cells are being studied in so much that the technique used to induce synchrony must not distort the basic metabolism of the cell. For this reason techniques involving the isolation of cells at a particular stage in the cell cycle are preferred to the addition of specific nutrients or inhibitors as a method of inducing cell synchrony. *Schizosaccharomyces pombe* is a valuable organism for the study of cell synchrony since growth results in an increase in the length of the rod-shaped cell and this can be used as a marker of the cell cycle. The onset of cell division is also marked by the formation of a transverse septum across the middle of the cell, and the appearance of this septum can be used to estimate the frequency of cells at the division stage of the cell cycle in the same way that the mitotic index is used in other eukaryotes. In *Saccharomyces cerevisiae* some indication of the stage of the cell cycle can be obtained from the size of the bud.

Studies on synchronously growing cells show that the increase in cell numbers occurs in a stepwise manner whereas the increase in the mass of the cells, indicated by the absorption of the cells in Fig. 5.20, is a continuous process. Biochemical studies have indicated that many cellular compounds are synthesized in a "step" or "periodic" manner while others increase continuously in a linear or exponential manner (Fig. 5.21). The most obvious cellular component synthesized periodically is DNA. Although both DNA synthesis and mitosis are essential for

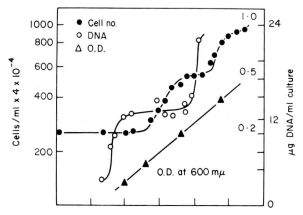

Fig. 5.20. Increase in cell numbers, cell mass (O.D.) and DNA levels in synchronous cultures of *Saccharomyces lactis*. From Tauro *et al.* (1969).

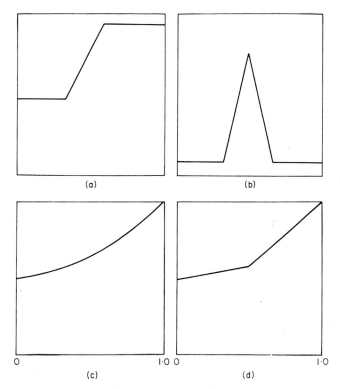

Fig. 5.21. Patterns of enzyme synthesis in synchronous cultures during one cell cycle: (*a*) step, (*b*) peak, (*c*) continuous exponential, (*d*) continuous linear. From Mitchison (1971).

cell division in the eukaryotic cell, they do not occur at the same time in the cell cycle. As has been previously discussed (Chapter 3) the eukaryotic cell cycle can be divided into four phases: G1, S, G2 and M phases. DNA synthesis occurs in the S phase and mitosis in the M phase, while G1 and G2 phases are respectively the periods before and after DNA synthesis. The times taken by the G and S phases in *Schizosaccharomyces pombe* are shown in Fig. 5.22.

Studies on the level of mitochondrial DNA in *Saccharomyces lactis* indicate that this is also synthesized in a synchronous manner but at a slightly different time in the cell cycle from nuclear DNA (Tauro *et al.*, 1969). In contrast to DNA, bulk RNA synthesis occurs continuously throughout the cell cycle. Using sucrose density gradients to separate out different RNA fractions, and isotope techniques to measure the rates of synthesis it has been shown that the large (26 S) and the small

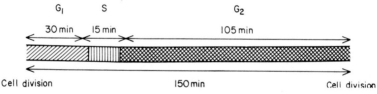

Fig. 5.22. Cell cycle of *Schizosaccharomyces pombe*.

(18 S) ribosomal subunits and tRNA are each synthesized continuously throughout the cell cycle (Tauro *et al.*, 1969).

In yeast, enzymes which are synthesized in a periodic and in a continuous linear manner have been described. In a synchronous culture, sucrase and acid and alkaline phosphatases increase linearly throughout the first cell cycle at a rate which is proportional to the cell numbers. Since there are twice as many cells per unit volume of culture during the second cycle the enzyme activity increases at twice the rate when measured on a volume basis (Fig. 5.23). The change in the rate of enzyme accretion does not occur at the time of cell division but approximately a quarter of the way through the second cell cycle. The time of the change is the same for each of the three enzymes.

A larger number of enzymes have been shown to be synthesized periodically, e.g. aspartate transcarbamylase, tryptophan synthetase and alcohol dehydrogenase (Mitchison, 1973). The level of each enzyme increases rapidly over a short period then remains constant throughout the remainder of the cell cycle. The time during the cell cycle when synthesis of a particular enzyme occurs is a characteristic of that enzyme. This constancy holds true for both constitutive and induced synthesis. In yeast the enzyme α-glucosidase can occur at

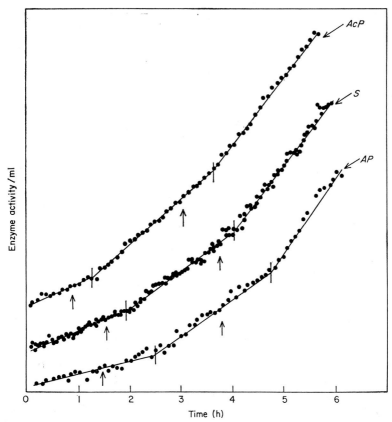

Fig. 5.23. Continuous linear enzyme synthesis in synchronous cultures of *Schizosaccharomyces pombe*.

Acid phosphatase	AcP
Alkaline phosphatase	AP
Sucrase	S

Arrows indicate the point of maximum cell-plate formation. Vertical lines indicate the point of change in the rate of enzyme synthesis. From Mitchison and Creanor (1969).

either a low basal level or a higher induced level. The enzyme is synthesized at exactly the same time in the cell cycle when the cells are grown in either the constitutive or the induced state (Fig. 5.24).

REGULATION OF THE BIOSYNTHESIS OF MACROMOLECULES

In an elegant series of experiments using a series of α-glucosidase mutants in *Saccharomyces cerevisiae* a correlation between the time of synthesis of a periodic enzyme and its position on the chromosome has

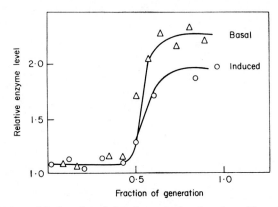

Fig. 5.24. Timing of induced and basal synthesis of α-glucosidase in synchronous cultures of *Saccharomyces cerevisiae*. From Halvorson *et al.* (1966).

been shown (Halvorson, 1966). In *S. cerevisiae*, six non-allelic structural genes for α-glucosidase (M1–6) have been described. Using strains containing these mutants it was demonstrated that in polyploid cultures, multiple allelic copies of the same structural genes, e.g. in an M_1M_1 diploid, are expressed at the same time whereas non-allelic structural genes for the same enzyme, e.g. in an M_1m_3 or m_1M_3 diploid, are expressed at different times (Fig. 5.25). The relationship between the period of expression of a gene and its position on the chromosome has been shown more clearly by Cox and Gilbert (1970). Two genes, gal_1 (for galactokinase) and lys_2 (for α-aminoadipic acid reductase) are closely linked in strain S288C of *S. cerevisiae* but further apart in strain 6 1009. The results shown in Fig. 5.26 show the increase in the interval between the times of synthesis of the two enzymes in the two strains. Such results suggest that the chromosome is read in a linear manner and that the DNA-RNA polymerase responsible for mRNA synthesis moves along the chromosome only once during each cell cycle such that each structural gene is transcribed at a specific time in the cell cycle. The lack of correlation between the time of gene expression and its distance from the centromere suggests that the DNA polymerase must move from one end of the chromosome to the other rather than from the centromere outwards if the model is correct (Tauro *et al.*, 1969; Fig. 5.27).

Although rRNA appears to be produced in a linear manner it is possible that the actual mechanism of synthesis is not too different from that of mRNA of periodic enzymes. It has been shown that if several genes for a single periodic enzyme are scattered at different loci on the chromosomes then the enzyme is synthesized in a series of

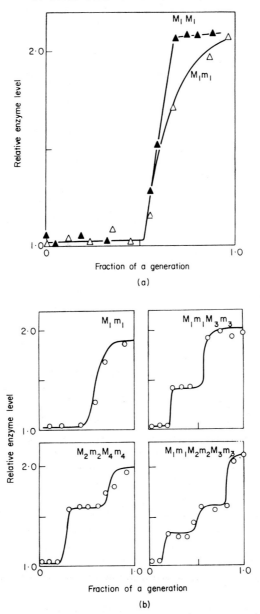

Fig. 5.25. Timing of synthesis of α-glucosidase in *Saccharomyces cerevisiae*.
(*a*) Diploid strains containing one and two allelic copies of the M genes. From Halvorson *et al.* (1964).
(*b*) Diploid strains containing one or more non-allelic copies of the M gene. From Tauro *et al.* (1969).

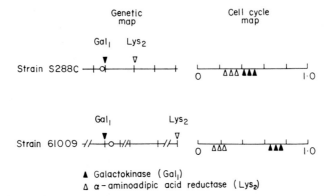

▲ Galactokinase (Gal₁)
△ α−aminoadipic acid reductase (Lys₂)

Fig. 5.26. Relationship between the time of appearance of two-step enzymes, galactokinase and α-aminoadipic acid reductase and their position on the chromosome in different strains of *Saccharomyces cerevisiae*. From Cox and Gilbert (1970).

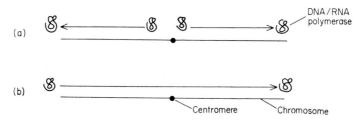

Fig. 5.27 Two models for the method of movement of DNA dependent RNA polymerase along the chromosome in yeast.

   (*a*) Movement of the polymerase outwards from the centromere would give a direct correlation between the time of expression of a gene and its distance from the centromere.

   (*b*) Movement of the polymerase from one end of the chromosome to the other would give no such correlation.

steps corresponding to each gene. It is evident that if a large number of genes were involved each step would be so small as to remain undetected and synthesis would appear to be continuous. DNA-RNA hybridization studies have shown that there are approximately 140 cistrons for each of the 18 S and 26 S rRNA subunits in the yeast genome. If these are scattered throughout the genome then apparent continuous synthesis would be expected. Since there are at least sixty different tRNA molecules in a yeast cell the linear production of tRNA could be explained in the same way. One of the most attractive aspects of this theory of the mechanism of synthesis of periodic enzymes is that it provides a biochemical rationale for the position of genes on chromosomes. The mechanism of control of the biosynthesis of linear enzymes is not clear, but it seems unlikely that it is the result of multiple structural genes (Mitchison, 1973). The difference between continuous and periodic enzymes may reflect post transcriptional events rather than a difference in the timing of mRNA production.

CONTROL OF THE CELL CYCLE AND ITS RELEVANCE TO
DIFFERENTIATION

Just as it has been possible in the past to divide the cell cycle into phases based on cytological criteria such as bud initiation and cell division, it is now possible to identify biochemical markers which occur at specific times in the cell cycle, such as periodic enzyme synthesis. A further set of markers which have not been mentioned are transition points, which are the points beyond which an inhibitor is no longer capable of affecting that cell cycle. Using these markers it is possible to map out a sequence of events which occur in a normal cell cycle. Using the known markers, attempts have been made to establish the causal relationships between different events and their relative importance in controlling the cell cycle. Mitchison (1973) has obtained evidence that there are at least two separate regulatory sequences in the cell cycle of *Schizosaccharomyces pombe*, the DNA-division cycle and the growth cycle. The DNA-division cycle contains the beginning and end of DNA synthesis, mitosis and cell division as markers, while the growth cycle is indicated by RNA synthesis, the synthesis of periodic enzymes and the action of certain inhibitors.

Using several inhibitors of DNA synthesis and cell division it has been possible to alter the speed of the DNA division cycle relative to the growth cycle. The addition of hydroxyurea to cells synchronized by a selection technique delays the S phase of the subsequent cycle for up to 2·5 h but has no effect on the time of synthesis of the periodic enzymes, ornithine and aspartate transcarbamylase and alcohol dehydrogenase. Similarly, mitomycin C which blocks nuclear division has no effect on the synthesis of periodic enzymes. In each case inhibition of the DNA division cycle has had no effect on the marker in the growth cycle. Comparable results have been obtained by inhibiting DNA synthesis using deoxyadenosine which acts at the beginning of the S phase. When deoxyadenosine is removed, one synchronous division occurs within half an hour and a second within 1·5 h. The division cycle has been condensed to 60% of its normal length. Cytological observations on the timing of nuclear division and cell plate formation indicate that the DNA division cycle has been accelerated rather than distorted.

These two divisions give rise to small cells because the growth cycle does not appear to be similarly accelerated. Periodic enzymes such as ornithine transcarbamylase are produced after the normal time lapse and the rate of RNA synthesis is not affected by the deoxyadenosine treatment.

If we consider differentiation to be concerned with the differential expression of the genome under different conditions and at different times, then a detailed knowledge of the control of the cell cycle should contribute to a greater understanding of the control of differentiation. Studies on periodic enzyme synthesis provide a mechanism for the differential expression of genes over a period of time and the separation of the control of DNA division and growth cycles indicates a possible mechanism for controlling the size and shape of cells.

5.6   Cell-Wall Construction and Reproduction

The foregoing studies present and to some extent prove the working hypothesis that a given cell morphology is determined by, and is dependent upon the chemical composition of the cell wall. In this respect, it is significant that differences exist in cell-wall structure of different   morphological variants of one fungus (McMurrough and Rose, 1967; Wang and Miles, 1966). A consideration of the factors involved in dimorphism could be profitable to a study of sporulation since asexual sporulation resembles the change from a mycelial to a yeast form in that both represent a change from a cylindrical to a spherical form of growth (see Chapter 6).

The differentiation of spores may not be immediately akin to Y formation, because Robertson (1966) concluded that the conidial rounding off in phialide formers is not a simple surface tension effect. Although they demonstrated a qualitative effect of cysteine on the morphology and cell-wall composition of *Neurospora crassa*, Bianchi and Turian (1967) stressed that the interconversion of mycelial (M) and conidial (C) forms of *N. crassa* is only apparently homologous otherwise this is a contradiction in terms to yeast ⇌ mould dimorphism.

Morton (1961) has attempted to relate sporulation and cell-wall composition in *Penicillium*. In his elegant experiments on the induction of sporulation he concluded that the most significant stimulation to sporulation is the emergence of submerged hyphae into the air, and he presents evidence to support the hypothesis that the aerial stimulus acts on the cell surface. The only difference between aerial and submerged hyphae is the sudden formation of an air/water interface, and it is possible that this could cause changes in the orientation of polar molecules at the cell surface such as the folding of protein. Weiss and Turian (1966) are of the opinion that the nitrogen source affects sporulation of *Neurospora crassa* through influencing the production or the physical state of some cell-wall component which is essential for the hyphae to break through the surface of the liquid medium and form conidia.

Thomas and Mullins (1967) have shown that the induction of branching in species of *Achlya* by substrate or by sexual hormones is accompanied by increased cellulose activity and probably depends upon wall softening. Wall softening by glucanases may prove to be a fundamental morphogenetic process in a wide variety of fungi (Bartnicki-Garcia, 1968a). In *Schizophyllum commune*, Wessels (1965) observed that pileus morphogenesis was correlated with considerable increases in the level of a specific repressible glucanase which hydrolyses the alkali-insoluble glucan of mycelial walls.

In conclusion, it can be said that the mechanism of cell-wall construction plays a decisive role in fungal morphogenesis. The current sophisticated investigations using chemical, enzymological and cytological techniques are gradually uncovering some of the biochemical bases of ontogenetic and phylogenetic development of the fungi.

Summary

One of the most intriguing problems in mycology is the understanding of how hyphal tube growth is established and maintained. It has long been known that hyphal fungi grow by apical deposition of new wall material although the mechanism of this process is only now being elucidated. To fully understand the operation of the biochemical machinery of apical growth it will be essential to know where cell-wall polymers are made and in particular to know where polymerization and/or assembly takes place. Are these wall polysaccharides synthesized somewhere in the endomembrane and then transported to the final site in the wall by special vesicles or are they predominantly assembled *in situ*? Current research would suggest that the microfibrillar network is assembled and perhaps polymerized *in situ* while the matrix material is preformed elsewhere, transported to the apex, and anchored to the wall. There is much circumstantial evidence for the existence of a high potential of synthetic and lytic activity in the apex of growing fungi and it is further suggested that these enzymes are carried to the apex in vesicles. Are these enzymes transported in the same or different vesicles and what controls the relative ratio of synthesis and migration? These important questions still remain unanswered. Furthermore, how do these vesicles move to the apex? There is no evidence that the vesicles are guided by microtubules or microfibrils and since they have no obvious means of propulsion it has been suggested that they migrate by electrophoresis. The existence of electrical potential in fungi has been established and this could generate a current strong enough to allow the electrophoretic movement of the vesicles.

During fungal growth, wall synthesis in positions other than the apex can cause a morphogenetic change of the hypha leading to septum and/or branch formation. In branch formation, the new outgrowth need not necessarily be similar in form or function (see Chapter 6 and 7) to the parent hypha. The control of subapical cell-wall synthesis is of paramount importance since it can often, under certain environmental conditions, lead to complex morphogenetic developments. Fruitful speculation must await a better understanding of the biochemical and biophysical steps in the synthesis of hyphal growth.

During apical growth in filamentous fungi, and bud formation in yeast cell-walls synthesis always takes place in an area where cell-wall material already exists. The importance of the existing cell-wall in controlling the architecture of the newly synthesized cell-wall is apparent from studies on wall regeneration in yeast protoplasts which were unable to synthesize a normal cell-wall. What is the role of the old cell-wall? Can new wall polymers only be incorporated on to an existing framework? Protoplast and spheroplast regeneration studies should ultimately supply some answers to these vexing problems.

The relationship between cell growth and replication and the control of cellular morphology is clearly illustrated by studies on yeast development. The recognition of the DNA-division cycle and the growth cycle as independently controlled processes within the cell provides a potential mechanism for the control of cell size. The separation of cell division from growth and nuclear division, which can be induced in species which exhibit yeast/mould dimorphism provides an ideal system for studies on the control of cell division and on the development of the filamentous growth form. The mechanism by which cell-wall synthesis is localized in yeast has much in common with those morphogenetic developments that will be discussed in Chapter 6.

The periodic synthesis of enzymes and the evidence that the chromosomal DNA is only transcribed once during each cell cycle in yeast presents fascinating problems as to the mechanisms of enzyme induction in yeast. If multiple copies of mRNA are required for enzyme induction a control mechanism must exist which ensures that each copy becomes active at the same time in the cell cycle. The sequential production of enzymes during the cell cycle must result in a continually changing enzyme composition. The possible effect of this on the metabolic control of the cell deserves further attention.

The foregoing studies permit and to some extent prove the working hypothesis that a given cell morphology is determined by, and is dependent upon the chemical composition of the cell-wall.

Recommended Literature

General Reviews

Aronson, J. M. (1965). The cell wall. *In* "The Fungi", **1,** 49–76. Ed. Ainsworth, G. C. and Sussman, A. S.   Academic Press, New York and London.

Bartnicki-Garcia, S. (1968a). Cell wall chemistry, morphogenesis and taxonomy in fungi. *Annual Review of Microbiology* **22,** 87–108.

Bartnicki-Garcia, S. (1970). Cell wall composition and other biochemical markers in fungal phylogeny. *In* "Phytochemical Phylogeny" pp. 81–103. Ed. Hasborne, J. B. Academic Press, New York and London.

Bartnicki-Garcia, S. (1973). Fundamental aspects of hyphal morphogenesis. *Symposium Society of General Microbiology* **23,** 245–267.

Bartnicki-Garcia, S. and McMurrough, I. (1971). Biochemistry of morphogenesis. *In* "The Yeasts", **2,** 441–491. Ed. Rose, A. H. and Harrison, J. S. Academic Press, London and New York.

Butler, G. M. (1966). Vegetative structures. *In* "The Fungi", **11,** 83–112. Ed. Ainsworth, G. C. and Sussman, A. S. Academic Press, New York and London.

Moore, R. T. (1965). The ultrastructure of fungal cells. *In* "The Fungi", **1,** 95–118. Ed. Ainsworth, G. C. and Sussman, A. S. Academic Press, New York and London.

Robertson, N. F. (1965). The mechanism of cellular extension and branching. *In* "The Fungi", **11,** 613–623. Ed. Ainsworth, G. C. and Sussman, A. S. Academic Press, New York and London.

Robertson, N. F. (1968). The growth process in fungi. *Annual Review of Phytopathology* **6,** 115–136.

Hyphal Differentiation

Bartnicki-Garcia, S. (1969). Cell wall differentiation in the Phycomycetes. *Phytopathology* **59,** 1065–1071.

Bartnicki-Garcia, S. and Lippman, E. (1969). Fungal morphogenesis: cell wall construction in *Mucor rouxii*. *Science* **165,** 302–304.

Bartnicki-Garcia, S. and Lippman, E. (1972). The bursting tendency of hyphal tips of fungi: presumptive evidence for a delicate balance between wall synthesis and wall lysis in apical growth. *Journal of General Microbiology* **73,** 487–500.

Bartnicki-Garcia, S. and Lippman, E. (1972). Inhibition of *Mucor rouxii* by polyoxin D: Effects of chitin synthetase and cell development. *Journal of General Microbiology* **71,** 301–309.

Bartnicki-Garcia, S., Nelson, N. and Cota-Robles, E. (1968). Electron microscopy of spore germination and cell wall formation in *Mucor rouxii*. *Archiv für Mikrobiologie* **63,** 242–255.

Brunswik, H. (1924). Untersuchungen uber die Geschlechtsund Kesnyerhaltnisse bie den Hymenomyceten gattung *Coprinus*. *In* "Botanische Abhandlungen" **5,** Ed. Goebel, K., 1–152. Gustav Fisher, Jena.

Girbardt, M. (1957). Der Spitzenkörper von *Polystictus versicolor* (L). *Planta* **50**, 47–59.

Girbardt, M. (1969). Die Ultrastruktur der Apikalregion von Pilzhyphen. *Protoplasma* **67**, 413–441.

Gooday, G. W. (1971). An autoradiographic study of hyphal growth of some fungi. *Journal of General Microbiology* **67**, 125–133.

Grove, S. N. and Bracker, C. E. (1970). Protoplasmic organization of hyphal tips among fungi: vesicles and spitzenkörper. *Journal of Bacteriology* **104**, 989–1009.

Grove, S. N., Bracker, C. E. and Morré, D. J. (1970). An ultrastructural basis for hyphal tip growth in *Pythium ultimum*. *American Journal of Botany* **57**, 245–266.

Hunsley, D. and Burnett, J. H. (1968). Dimensions of microfibrillar elements in fungal walls. *Nature, London* **218**, 462–463.

Hunsley, D. and Burnett, J. H. (1970). The ultrastructural architecture of the walls of some hyphal fungi. *Journal of General Microbiology* **62**, 203–218.

Katz, D. and Rosenberger, R. F. (1971). Hyphal wall synthesis in *Aspergillus nidulans*: Effect of protein synthesis inhibition and osmotic shock on chitin insertion and morphogenesis. *Journal of Bacteriology* **108**, 184–190.

Marchant, R. and Smith, D. G. (1968). A serological investigation of hyphal growth in *Fusarium culmorum*. *Archiv für Mikrobiologie* **63**, 85–94.

McClure, K. W., Park, D. and Robinson, P. M. (1968). Apical organization in the somatic hyphae of fungi. *Journal of General Microbiology* **50**, 177–182.

McMurrough, I., Flores-Carreon, A. and Bartnicki-Garcia, S. (1971). Pathway of chitin synthesis and cellular localization of chitin synthetase in *Mucor rouxii*. *Journal of Biological Chemistry* **246**, 3999–4007.

Nishi, A., Yanagita, T. and Maruyama, Y. (1968). Cellular events occurring in growing of *Aspergillus oryzae* as studied by autoradiography. *Journal of General and Applied Microbiology* **14**, 171–182.

Park, D. and Robinson, P. M. (1966). Internal pressure of hyphal tips of fungi and its significance in morphogenesis. *Annals of Botany* **30**, 425–439.

Peberdy, J. F. (1972). Protoplasts from fungi. *Science Progress* **60**, 73–86.

Vegetative Morphogenesis

Galbraith, J. C. and Smith, J. E. (1969). Filamentous growth of *Aspergillus niger* in submerged shake culture. *Transactions of the British Mycological Society* **52**, 237–246.

Pirt, S. J. (1966). A theory of the mode of growth of fungi in the form of pellets in submerged culture. *Proceedings of the Royal Society B* **166**, 369–373.

Righelato, R. C. (1974). Growth kinetics of mycelial fungi. *In* "The Filamentous Fungi", *Industrial Mycology*. Ed. Smith J. E. and Berry, D. R. Edward Arnold (Publishers) Ltd., London.

Trinci, A. P. J. (1970). Kinetics of the growth of mycelial pellets of *Aspergillus nidulans*. *Archiv für Mikrobiologie* **73**, 353–367.

Trinci, A. P. J. (1971). Influence of the width of the peripheral growth zone on the radial growth rate of fungal colonies on solid media. *Journal of General Microbiology* **67**, 325–344.

Yanagita, T. and Kogane, F. (1962). Growth and cytochemical differentiation of mold colonies. *Journal of General and Applied Microbiology* **8**, 201–213.

Vegetative Differentiation in Yeasts

Barker, J. T., Srb, A. M. and Steward, F. C. (1969). Protein morphology and genetics in *Neurospora*. *Developmental Biology* **20**, 105–124.

Berry, E. A. and Bevan, E. A. (1972). A new species of double stranded RNA from yeast. *Nature, London* **239**, 279–280.

Brody, S. (1970). Correlation between reduced NADPH levels and morphological change in *Neurospora crassa*. *Journal of Bacteriology* **101**, 802–807.

Brown, C. M. and Hough, J. S. (1965). Elongation of yeast cells in continuous culture. *Nature, London* **206**, 676–678.

Cox, C. G. and Gilbert, J. B. (1970). Non-identical times of gene expression in two strains of *Saccharomyces cerevisiae* with mapping differences. *Biochemical and Biophysical Research Communications* **38**, 750–757.

Esser, K. and Kuenen, R. (1967). "Cytogenetics of Fungi". Springer Verlag, Berlin.

Halvorson, H. O., Carter, B. L. A. and Tauro, P. (1971). Synthesis of enzymes during the cell cycle. *Advances in Microbial Physiology* **6**, 47–106.

Halvorson, H. O., Gorman, J., Tauro, P., Epstein, R. and LaBege, M. (1964). Control of enzyme synthesis in synchronous cultures of yeast, *Federation Proceedings* **23**, 1002–1008.

Halvorson, H. O., Brock, R. M., Tauro, P., Epstein, R. and LaBerge, M. (1966). Periodic enzyme synthesis in synchronous cultures of yeast. *In* "Cell Synchrony", pp. 102–116. Ed. Cameron, I. L. and Padilla, G. M. Academic Press, London and New York.

Jinks, J. C. (1958). Cytoplasmic differentiation in fungi. *Proceedings of the Royal Society B* **148**, 314–321.

Kreger-Van Rij, N. J. W. (1969). Taxonomy and systematics of yeasts. *In* "The Yeasts", **1**, 8–78. Ed. Rose, A. H. and Harrison, J. S. Academic Press, London and New York.

Lodder, J. (1970). ed. "The Yeasts." North Holland, Amsterdam.

Mahadewan, P. R. and Tatum, E. L. (1965). The relationship of the major constituents of the *Neurospora crassa* cell wall to wild type and colonial morphology. *Journal of Bacteriology* **90**, 1073–1081.

Misra, N. C. and Tatum, E. L. (1970). Phosphoglucomutase mutants of *Neurospora sitophila* and their relation to morphology. *Proceedings of the National Academy of Sciences U.S.* **66**, 628–645.

Mitchison, J. M. (1971). "The biology of the cell cycle". Cambridge University Press.

Mitchison, J. M. (1973). The cell cycle of a eukaryote. *Symposium Society for General Microbiology* **23**, 189–208.

Mitchison, J. M. and Creanor, J. (1969). Linear synthesis of sucrose and phosphatase during the cell cycle of *Schizosacchasomyces pombe*. *Journal of Cell Science*, **5**, 373–391.

Nickerson, W. J. (1963). Molecular basis of form in yeast. *Bacteriological Reviews* **27**, 305–324.

Pfaff, H. J. (1971). Structure and biosynthesis of the yeast envelope. *In* "The Yeasts", **11**, pp. 135–210. Ed. Rose A. H. and Harrison, J. S. Academic Press, London and New York.

Scott, N. C. and Tatum, E. L. (1970). Glucose-6-phosphate dehydrogenase and *Neurospora* morphology. *Proceedings of the National Academy of Sciences U.S.* **66**, 515–522.

Sentheshanmuganathan, S. and Nickerson, W. J. (1962). Composition of cells and cell walls of triangular and ellipsoidal forms of *Trigonopsis v variables*, *Journal of General Microbiology*, **27**. 451–464.

Shimoda, C. and Yanagishima, N. (1968). Strain dependence of the cell expanding effect of β1, 3 glucanase in yeast. *Physiologia Plantarum* **21**, 1163–1169.

Tauro, P., Schweizer, E., Epstein, R. and Halvorson, H. O. (1969). Synthesis of macromolecules during the cell cycle in yeast. *In* "The Cell Cycle", 101–118. Ed. Padilla, G. M. Whitson, G. L. and Cameron, I. L. Academic Press, London and New York.

Turner, W. B. (1971). "Fungal Metabolites". Academic Press, London and New York.

Williamson, D H. (1966). Nuclear events in synchronously dividing yeast cultures. *In* "Cell Synchrony", pp. 81–101. Ed. Cameron, I. L. and Padilla, G. M. Academic Press, London and New York.

Yeast/Mould Dimorphism

Bartnicki-Garcia, S. (1963). Symposium on the biochemical bases of morphogenesis in fungi. III. Mold-yeast dimorphism in *Mucor*. *Bacteriological Reviews* **27**, 293–304.

Bartnicki-Garcia, S. and Nickerson, W. J. (1962a). Induction of yeast-like development in *Mucor* by carbon dioxide. *Journal of Bacteriology* **84**, 829–840.

Bartnicki-Garcia, S. and Nickerson, W. J. (1962b). Nutrition, growth, and morphogenesis of *Mucor rouxii*. *Journal of Bacteriology* **84**, 841–858.

Bartnicki-Garcia, S. (1968b). Control of dimorphism in *Mucor* by hexoses: inhibition of hyphal morphogenesis. *Journal of Bacteriology* **96**, 1586–1594.

Haidle, C. W. and Storck, R. (1966). Control of dimorphism in *Mucor rouxii. Journal of Bacteriology* **92**, 1236–1244.

Kanetsuna, F., Carbonell, L. M., Azuma, I. and Yamamura, Y. (1972). Biochemical studies on the thermal dimorphism of *Paracoccidioides brasiliensis. Journal of Bacteriology* **110**, 208–218.

Nickerson, W. J. and Falcone, G. (1956). Identification of protein disulfide reductase as a cellular division enzyme in yeasts. *Science* **124**, 722–723.

Romano, A. H. (1966). Dimorphism. *In* "The Fungi" **11**, 181–209. Ed. Ainsworth, G. C. and Sussman, A. S. Academic Press, New York and London.

Terenzi, H. F. and Storck, R. (1969). Stimulation of fermentation and yeast-like morphogenesis in *Mucor rouxii* by phenethyl alcohol. *Journal of Bacteriology* **97**, 1248–1261.

Warburg, O., Geissler, A. W. and Lorenz, S. (1968). Oxygen, the creator of differentiation. *In* "Aspects of Yeast Metabolism", 327–336. Ed. Mills, A. K. F. A. Davis Co, Philadelphia.

Cell-Wall Construction and Reproduction

Bianchi, D. E. and Turian, G. (1967). The effect of nitrogen source and cysteine on the morphology, conidiation and cell wall fraction of conidial and aconidial *Neurospora crassa. Zeitschrift für Mikrobiologie* **7**, 257–264,

McMurrough, I. and Rose, A. H. (1967). Effect of growth rate and substrate limitation on the composition and structure of the cell wall of *Saccharomyces cerevisiae. Biochemical Journal* **105**, 189–199.

Morton, A. G. (1961). The induction of sporulation in mould fungi. *Proceedings Royal Society B.* **153**, 548–564.

Thomas, D. S. D. and Mullins, J. T. (1967). Role of enzymatic wall softening in plant morphogenesis: hormonal induction in *Achlya. Science* **156**, 84–85.

Wang, C. S. and Miles, P. G. (1966). The physiological characterization of dikaryotic mycelia of *Schizophyllum commune. Physiologia Plantarum* **17**, 573–580.

Weiss, B. and Turian, G. (1966). A study of conidiation in *Neurospora crassa*. *Journal of General Microbiology* **44,** 407–412.

Wessels, J. G. H. (1965). Morphogenesis and biochemical processes in *Schizophyllum commune*. *Wentia* **13,** 1–113.

Terenzi, H. F. and Storck, R. (1969). Stimulation of fermentation and yeast-like morphogenesis in *Mucor rouxii* by phenethyl alcohol. *Journal of Bacteriology* **97,** 1248–1261.

6 Asexual Reproduction

6.1 Introduction

Although most fungi have the ability to reproduce by the formation of sexual spores (i.e. involving meiosis), all fungi can perpetuate themselves by asexual means. There exists within the fungi a vast range of asexual mechanisms producing limited numbers or large numbers of reproductive survival units depending on the fungus and on the environmental conditions. Many fungi have only one type of asexual reproduction while others have two (or more) which can develop in response to different environmental stimuli.

The classification of asexual reproductive mechanisms is difficult and controversial, and in the present context the system of subdivision proposed by Alexopoulos (1962) will be adopted. However, it is fully realized that a strong case can be made for certain types of asexual reproduction being considered in other sections, e.g. chlamydospores in Chapter 4 and sclerotia in either Chapter 5 or 7.

Fragmentation of hyphae. In this instance any portion of the somatic mycelium which becomes detached and is deposited on a suitable nutrient surface may continue to grow and produce a new mycelium genetically similar to the parent colony. In the *mycelia sterilia* this represents the only means of species perpetuation. However, almost all other filamentous fungi can be grown by fragmentation, and indeed this is the normal laboratory method of subculturing most filamentous fungi. In fermenter culture the mechanical action of the impeller blades often causes severance into fragments of mycelia which subsequently continue growing. The fungal hypha possesses remarkable properties of regrowth and may be cut into extremely small sections and yet still retain the ability to regenerate. In laboratory and industrial fermenter studies a mycelial inoculum is often cut into smaller and more numerous growth units by prior treatment in a Waring blender.

Chlamydospore formation. Chlamydospores are produced by many fungi to enable the fungus to survive unfavourable conditions such as low temperatures, lack of nutrients, etc. They may be terminal or inter-calary cells of a hypha or even single cells which enlarge and round up, store supplies of food, and form a thick resistant wall. Such cells may live for many years until favourable conditions return. Although they have nothing to do with the sexual stage of the fungus they are often likened to the long-lived sexual memnospores of certain fungi or to resistant bacterial endospores. However, structurally and functionally they are very similar to the spherules of the Myxomycetes.

Sclerotium formation. Sclerotia are firm aggregations of somatic hyphae with determinate growth. They contain numerous forms of reserve food materials and after maturation are capable of independent exist-ence for a considerable period of time. They may vary in size from a few cells to 10 cm across, and they often have a definite dark-coloured and usually hard outer rind enclosing a medulla of densely packed hyphae. The internal hyphae lack consistent orientation. Small sclerotia are often termed microsclerotia. Functionally, these sclerotia are similar to the plasmodial sclerotinization found in the Myxomycetes.

Fission. This method of propagation is normally found in certain uni-cellular fungi, principally yeasts. Firstly nuclear division occurs followed by cytokinesis and constriction to form two equal daughter cells. This method of propagation is widely distributed in other microbial forms, particularly the bacteria and blue-green algae.

Budding. In this mechanism a small outgrowth develops from a parent cell. As the bud or outgrowth forms, the nucleus of the parent cell divides and one daughter nucleus migrates into the bud. Budding is found in many yeasts but can also occur in many other fungi at certain phases of the life cycle or under highly exacting environmental condi-tions for growth.

Spore formation. Spores are normally thin-walled, unicellular, bicellular or multicellular units formed on or in the parent thallus and are released by various means into the environment. They are dispersed by wind or water and eventually settle down in a suitable environment and produce a germ tube which then grows into a vegetative thallus similar to the parent. This is by far the most common method of asexual propagation within the fungi.

In certain Phycomycetes asexual reproduction occurs by the for-mation, within a sporangium, of naked cells, zoospores, which upon release swim away by means of anteriorly, laterally or posteriorly attached flagella. Zoospores eventually settle down and become en-cysted, and this cyst becomes the start of the new thallus. However, in the majority of fungi including many Phycomycetes and all of the Higher Fungi in which asexual reproduction occurs, no motile spores

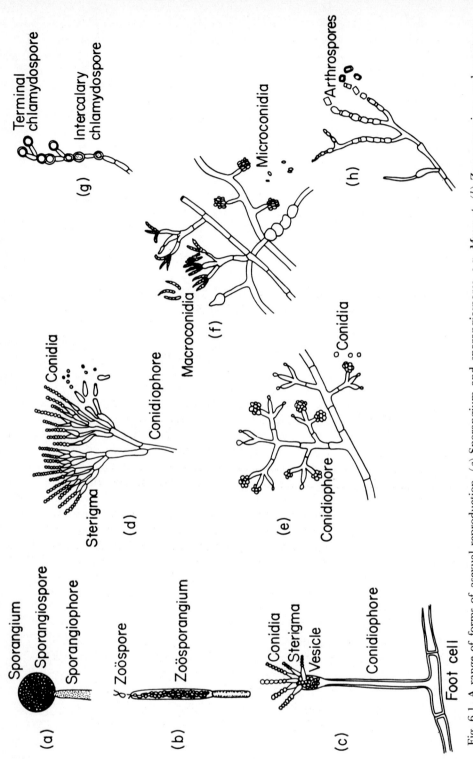

Fig. 6.1. A range of forms of asexual reproduction. (a) Sporangium and sporangiospores, e.g. *Mucor sp.* (b) Zoosporangium and zoospores, e.g. *Saprolegnia sp.* (c–e) Unicellular conidia. (c) *Aspergillus sp.*; (d) *Penicillium sp.*; (e) *Trichoderma sp.* (f) Macroconidia and clustered microconidia of *Fusarium*. (g) Chlamydospores of *Fusarium*. (h) Arthrospores, e.g. *Geotrichum sp.*

are produced but rather the spores are covered with a rigid wall and are usually passively distributed by air and water. These spores have several types of origin. In the Mucorales they are produced within a sporangium and upon the rupture or dissolution of the sporangial walls the spores are set free. Spores that arise as single, separable cells of the mycelium are called conidia. They may be formed by the fragmentation of the whole mycelium or of special hyphae into cylindrical, ovoid or spherical cells called oidia or arthrospores; or they may arise from the formation of terminal or lateral cells from special hyphae. Hyphae that bear conidia are called conidiophores. Such conidiophores may be morphologically very similar or very different from vegetative hyphae. Growth in length of conidiophores and their branching is usually restricted to the apical region and may or may not cease with the production of a terminal conidium or chain of conidia.

The manner in which conidia are attached to their conidiogenous cells and the mechanisms by which they are liberated are not only of biological interest but also provide characters useful for classification. A range of forms of asexual reproduction is shown in Fig. 6.1.

The Influence of Environmental Factors

The wealth of information on environmental factors which affect asexual reproduction has been competently summarized by Hawker (1966). Almost all environmental conditions can be shown to influence fungal development and, in particular, the effect of light, humidity, aeration, pH value, temperature and nutrient type and concentration have been extensively investigated. There is no obvious theme underlying the complex effect of environmental factors on development and the enormous array of facts merely substantiates the principles set forth in 1898 by George Klebs and re-enunciated in modern terminology by Morton (1967). "The conditions for vegetative growth and for reproduction are different. Some minimum period of vegetative development is required before the organism becomes competent to produce reproductive bodies, during which it synthesizes specific metabolites, enzymes or food substances essential for reproduction. Reproduction is often induced when some external or internal factor, frequently some nutrient, becomes limiting for vegetative development. The external conditions inducing reproduction are usually narrower in range and more specific than those permitting vegetative growth." Little is known of the mechanism by which perception of an external stimulus is translated into the visible initiation of the reproductive phase. Where are the receptor sites for environmental stimuli? Do physical and chemical

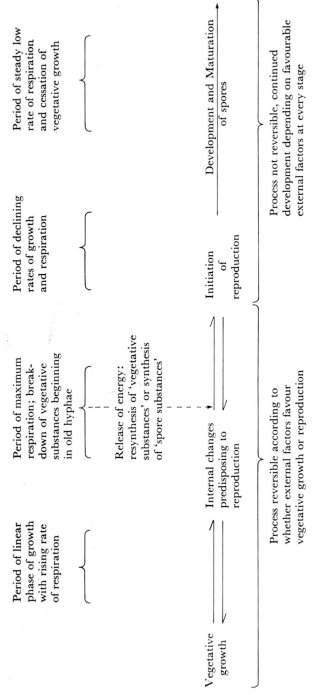

Fig. 6.2. Diagram to show suggested relation between growth, respiration, and sporulation in a fungus genetically competent to produce spores, growing on static culture. From Hawker and Hepden (1962).

stimuli exert their influence through similar mechanisms? Answers to these questions will only come with deeper understanding of the bio-chemical and biophysical properties of the cell together with whole organism studies.

The ability of a fungus to change from a vegetative growth form to a reproductive form is the result of a combination of genetic competence to produce the new form and the effect of environmental factors. A fungal mycelium which is genetically unable to undergo the change will remain sterile under all environmental conditions, mycelium with a low potential for differentiation will change only when the external conditions are particularly suitable, while mycelium with a high potential for differentiation may be prevented by sublethal environ-mental conditions. Fig. 6.2 shows a probable series of events during the change from vegetative growth to the reproductive phase.

6.2 Chlamydospore Development

Chlamydospores can be induced in many fungi by cultivation in low nutrient media, leaching of established agar cultures, incubation of ungerminated or germinated spores in salts solution, ageing of cultures, or co-culturing with specific bacteria or bacterial products. Several days are required for induction and formation of chlamydospores with all of these methods and consequently they are somewhat unsuitable for precise biochemical investigation. However, Cochrane and Cochrane (1971) using *Fusarium solani* have now developed a method by which un-germinated or germinated macrospores can be induced to form chlamy-dospores in relatively short time periods (Fig. 6.3). This system allows the study of the early events in this developmental process, in particular, the transcription and translation of the genetic information. A second advantage is that the system gives chlamydospores of good purity and high bulk yield.

When macrospores of *Fusarium solani* are incubated in a complete medium at pH 6·0 they germinate and produce a normal branching mycelium. Incubation of macrospores in a complete medium at pH 4·0 prevents normal germination and instead chlamydospores are formed in 18–24 h; pregerminated macrospores (spores which had been incubated in a complete medium at pH 6·0 for 10 h) when incubated in a complete medium at pH 4·0 form chlamydospores in 6–12 h. Maximum chlamy-dospore induction from ungerminated spores requires carbon, nitrogen and mineral salts. Induction from pregerminated spores has a carbon requirement but no requirement for nitrogen or mineral salts. The process of chlamydospore induction in ungerminated spores is

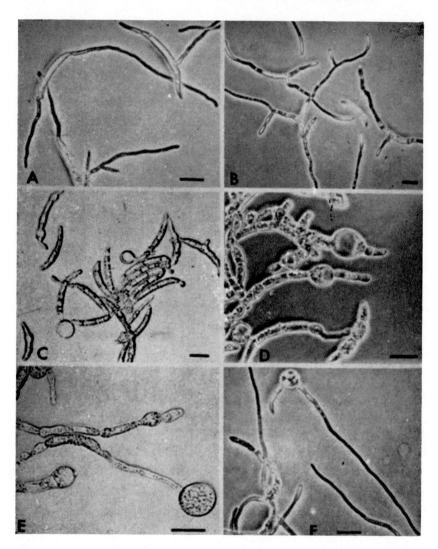

Fig. 6.3. Germination and chlamydospore formation in *Fusarium solani*. Bar equals 20 μm. A. Germinating macrospores after 10 h in basal pH 6·5 medium (phase contrast). B. New macrospores forming on germ tubes after 17 h in basal pH 6·5 medium. C. Formation of chlamydospore initials by macrospores after 10 h in pH 4·0 medium (bright field). D. As C, after 24 h in pH 4·0 medium (phase contrast). E. Developing chlamydospore on a germ tube; macrospores were germinated 10 h in basal pH 6·5 medium, then incubated 24 h in pH 4·0 medium (bright field). F. Germinating chlamydospore; macrospores were incubated 36 h in pH 4·0 medium, and then transferred for 12 h to basal pH 6·5 medium (phase contrast). From Cochrane and Cochrane (1971).

aerobic but can occur under anaerobic conditions for pregerminated spores.

The specific nutritional requirements of the ungerminated spore →chlamydospore transition are contrary to the hypothesis that this process occurs simply in response to starvation. Rather, it implies that specific substances, in this case mimicked by H^+, are responsible for chlamydospore initiation. The failure of pregerminated spores to require nitrogen and mineral salts for chlamydospore formation is probably due to previous assimilation during germination and subsequent storage.

Pregerminated macrospores can still form chlamydospores in the presence of levels of cycloheximide which inhibit protein synthesis as well as almost 99% of the incorporation of uracil ^{14}C into ribonucleic acid. Ungerminated macrospores cannot form chlamydospores in the presence of similar levels of cycloheximide. Time course experiments have shown that the information for the formation of chlamydospores appears early in normal macrospore germination, and is only expressed if the environment becomes unsuitable for normal germination to proceed. Thus, the germinated macrospore has a preformed system for making a chlamydospore since a total inhibition of protein synthesis and concurrent inhibition of RNA synthesis fails to influence either the rate of chlamydospore formation or the extent of final conversion (Cochrane and Cochrane, 1970). If an act of morphological differentiation requires prior transcription and translation of a genetic message it must be assumed that both of these processes have been completed in the germinated macrospore. Transcription and translation in chlamydospore formation from germinated macrospores clearly precedes the development of the visible morphological process. Another example of this phenomenon occurs during germination of *Blastocladiella* spores (see p. 100). How the transcribed and translated messages exist in a dormant state is not known. Perhaps such a situation may exist in other types of development.

Chlamydospore formation can be considered to have ecological significance since it allows the fungus to form a resistant structure under conditions which limit or inhibit macromolecular synthesis. As an experimental system chlamydospore formation warrants further study. The resemblance between this process and spherulation in *Physarum* is particularly apparent, and similar experimental investigations could well lead to a better understanding of wall chemistry in true fungi.

6.3   Sclerotium Development

Sclerotium formation is a complex morphogenetic process in which both structural and biochemical events take place. In most cases sclerotia arise as discrete initials among the somatic mycelium. In terminal formation the initial is formed by condensed terminal growth and extensive branching of an individual hyphal tip. Examples of terminal sclerotium formation occur in *Botrytis cinerea* and *Pyronema domesticum*. A second type of sclerotium formation occurs in *Sclerotinia gladioli* and *Verticillium dahliae*. Intercalary segments of one of more hyphae develop by the formation of additional septa and the production of numerous side branches or budlike outgrowths (Fig. 6.4). In each case after initiation there is an increase in size by extensive branching, coiling and interweaving of the hyphae. At maturation there is marked delimitation and internal consolidation. Late in development the reserve food material accumulates either as wall thickening or as cell lumina. Within each sclerotium there is multiple zonation, and regrowth in suitable conditions occurs at localized but random points on the medulla.

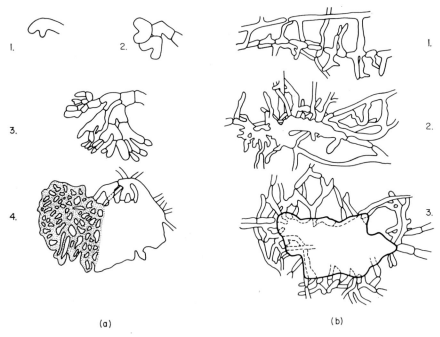

(a)   (b)

Fig. 6.4. Sclerotium development. (*a*) Stages in the development of *Botrytis allii* (terminal type). ×225. (*b*) Stages in development of *Sclerotinia gladioli* (strand type). ×225. From Townsend and Willetts (1954).

It has been considered that the formation of sclerotia by fungi in culture is induced by the accumulation of staling products in the medium. In addition zinc may be important for their formation and the length of the photoperiod may determine their size and number. However, the biochemical events which lead to the initiation of sclerotia remain almost completely unknown.

By using metabolic inhibitors, chelating compounds and sulphur-containing amino acids Chet and Henis (1968) have suggested that during growth of *Sclerotium rolfsii* sclerotium formation is repressed by a protein containing copper and a sulphydryl group, the effects of which can be artificially removed by the presence of potassium iodate, iodo-acetic acid or Na_2EDTA. However, recent studies would suggest that the effects are indirect in action. Rather than affecting morphogenesis directly they would appear to alter unrelated metabolic processes and these changes are then reflected in the number of sclerotia subsequently produced.

In *Sclerotinia sclerotiorum* and *S. trifoliorum* there are three phenomena during the early phases of sclerotium development: the formation of liquid droplets containing soluble carbohydrates on the surfaces of young sclerotia, a decrease in the water content, and a decrease in endogenous mannitol and also of glucose and trehalose in *S. trifoliorum*. The third phenomenon probably represents the conversion of soluble carbohydrates to storage and structural compounds, thus maintaining the movement of soluble carbohydrates into the developing sclerotia by creating a metabolic sink. Excretion of carbon compounds in droplets also has this effect.

Synchronous formation of sclerotia in *S. rolfsii* has recently been achieved by adding 0·5% (w/v) lactose to a synthetic medium containing glucose. Sclerotia are subsequently formed in a circular band at the colony margin. The ability to achieve a synchronous development of sclerotia will now allow a better understanding of the structural and biochemical changes preceding and accompanying this type of morphogenesis. How lactose can promote synchrony of sclerotial formation is not known. It is possible that lactose inhibits apical growth and encourages excessive branching. High osmotic pressure can cause extensive branching in *Aspergillus nidulans* (Chapter 5) by disturbing the pattern of wall synthesis. Autoradiographic incorporation studies may well show changes in the pattern of vesicle distribution and of wall synthesis in the presence of lactose. The complex but regular morphology of the sclerotium must further imply intercellular reactions, perhaps by physical contact and/or metabolite exchange between neighbouring cells. Again, do hormones become involved? Obvious similarities exist between sclerotium formation, and on the one hand fungal strand formation (Chapter 5) and on the other basidiocarp and ascocarp formation in higher fungi (Chapter 7).

Patterns of soluble proteins and enzymes have been examined in the vegetative mycelium of *Sclerotium rolfsii* and at several stages during synchronous sclerotium formation. By examining isozymes of several enzymes it has been shown that each developmental stage has a specific and distinctive pattern (Chet, Retig and Henis, 1972). Sclerotium formation undoubtedly represents a most challenging problem of multicellular development.

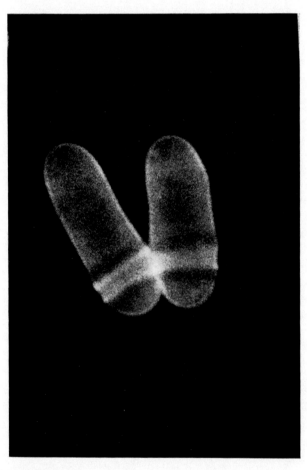

Fig. 6.5. Division scars in *Schizosaccharomyces pombe* (Streiblova, personal communication).

6.4 Propagation by Fission and Budding

Growth and vegetative reproduction in yeasts can be considered as a continuous increase in cell mass associated with cell division at regular intervals. In the fission yeasts growth occurs terminally, usually at one end of the cell, and division is achieved through the development of a double septum half-way along the tubular cell. Cell separation occurs when the septum is complete leaving it in the position of the end wall of the daughter cells, from which future growth occurs. The site at which the septum was formed can be recognized, even after another period of growth and cell division has occurred, as a dark band when the cells are examined by fluorescence microscopy (Fig. 6.5). Thus details of the cell architecture are a direct result of the method of cell division.

In yeasts which reproduce by budding, the site of synthesis of the bud can be recognized on the parent cell as a bud scar and on the daughter cell as a birth scar (Fig. 6.6). *Saccharomyces cerevisiae* has a multipolar method of budding. Buds can be produced at any point on the surface of the cell, except in a region where a bud scar derived from a previous

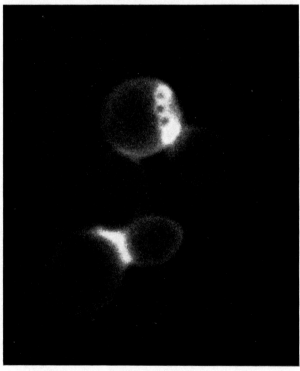

Fig. 6.6 (*a*)

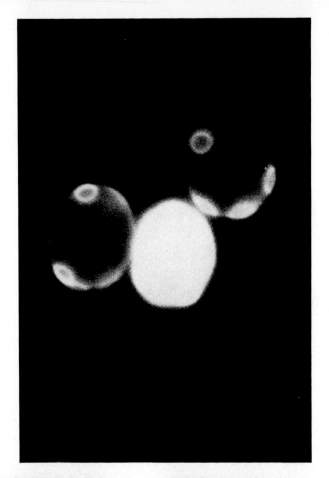

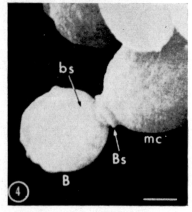

Fig. 6.6. (a) Bud scars in haploid *Saccharomyces cerevisiae*. (b) Bud scars in diploid *Saccharomyces cerevisiae*. (c) and (d) Scanning electron micrographs of budding cells of *Saccharomyces uvarum* (a) and (b), Streiblova, personal communication; (c) and (d), Belin (1972).

Bs = bud scar   B = bud
bs = birth scar   mc = mother cell

cell division is situated. Since no two buds arise from the same point the number of bud scars indicates the number of times the cell has been involved in cell division as the parent cell, and can be used as a measure of the physiological age of the cell. Correlation of the physiological age of cells with size has shown that a slight increase in size occurs with age and that this increase can be correlated with the increase in the area of cell wall material contributed by the bud scars (Mortimer and Johnston, 1959). One consequence of this method of division in which the products of the division are not identical is that the population of cells grown in the most carefully controlled conditions must always be heterogeneous with respect to age and cell size. Bipolar budding occurs

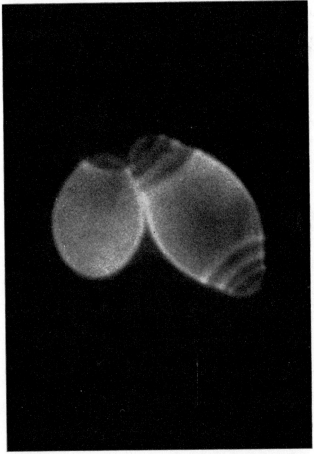

Fig. 6.7. Multiple concentric bud scars in *Saccharomyces ludwigii* (Streiblova, personal communication).

in *S. ludwigii*. Successive generations of buds arise at the same site on the cell wall resulting in a series of concentric bud scars (Fig. 6.7).

The site of bud formation in budding yeast can be recognized using several different techniques. An increase in the level of sulphydryl groups at the site of bud initiation can be detected using cytochemical assays. Associated with this, there occurs an increase in the level of protein disulphide reductase, the enzyme responsible for reducing disulphide bonds in the cell wall. The site of bud formation is also indicated by an accumulation of vesicles in the adjoining cytoplasm.

Bud formation always takes between 1 and 2 h which in cells growing rapidly occupies the whole of the cell cycle. However, if the cell cycle is extended by growing yeast in nutrient limiting conditions the budding time remains the same. Cell growth and cell replication appear to be independent processes which can be separated under controlled physiological conditions. Growth continues in a linear manner throughout the

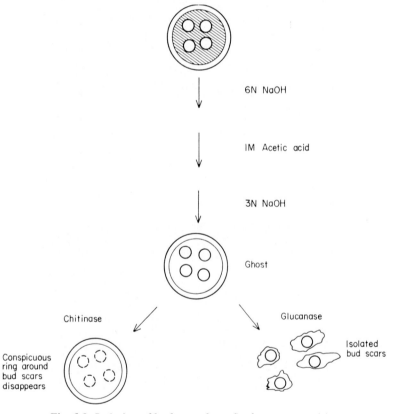

Fig. 6.8. Isolation of bud scars from *Saccharomyces cerevisiae*.

cell cycle whereas division occurs in a stepwise manner at a specific time in the cell cycle. The synthesis of DNA and many enzymes in yeast occurs in a similar stepwise manner (Halvorson, Carter and Tauro, 1971), but the synthesis of rRNA and some enzymes occurs in a linear manner (Mitchison, 1971).

The mechanism by which the bud is separated from the parent cell is not fully understood. It seems likely that it involves both physiological and physical processes. Physiological processes specific to cell separation as opposed to cell growth are indicated by the observation that buds fail to separate in yeast cultures grown in inositol or biotin-deficient media. This failure to separate is associated with abnormal cell-wall formation and an increase in the ratio of glucan to mannan in the cell-wall.

Studies on the biochemistry of bud scars indicate that chitin could play an important role in bud excision. Chitin has been shown to be located in the annular ring surrounding bud scars in *Saccharomyces carlsbergensis* (Cabib and Bowers, 1971). Ghosts of cells of *S. carlsbergensis* were prepared by extracting whole cells with alkali and acetic acid. These were then treated with purified preparations of glucanase and chitinase. Treatment with chitinase resulted in the removal of material from around the bud scars leaving the cell ghosts intact whereas treatment with glucanase destroyed the ghosts leaving isolated bud scars (Fig. 6.8).

The synthesis of chitin occurs during a short period of the cell cycle in

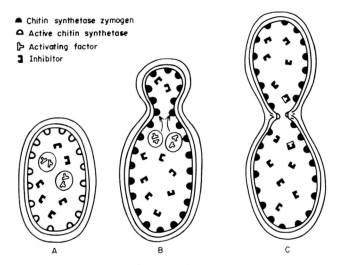

Fig. 6.9. Hypothetical scheme for the initiation of chitin synthesis (adapted from Cabib and Farkas, 1971).

synchronous yeast cultures. The main enzyme involved, chitin synthetase, has been shown to exist in the cell as an inactive zymogen which
can be activated by treatment with trypsin. An endogenous activating
enzyme has been isolated from yeast cells. This enzyme appears to be
subject to inhibition by a second, heat stable, protein which has been
isolated. A model for the control of chitin synthesis involving these three
components has been put forward by Cabib and Farkas (1971). They
postulated that the chitinase zymogen is present in the plasmalemma and
that the activating enzyme occurs within vesicles in the cytoplasm.
Synthesis occurs when the chitin zymogen is activated by direct contact
between the vesicles and the plasmalemma. The inhibitor is believed to
neutralize the activating enzyme once chitinase has been generated in
the specific region of bud formation (Fig. 6.9).

6.5   Spore Formation

Biogenesis of Endogenous Spores

In this type of asexual spore formation the protoplast of a single, walled
cell—the sporangium—is reorganized to form a number of smaller
cells. There is a considerable increase in the extent of cell surface and
concomitant increase in plasmalemma. Cytoplasmic division occurs
separately from nuclear division and cell-wall formation. In zoospore
formation naked, flagellated cells are produced whereas in sporangiospore formation a rigid cell is developed before the spore is discharged
into the environment. The biochemistry of zoospore biogenesis has been
extensively studied whereas sporangiospore formation has been mostly
neglected.

ZOOSPORES

The most meaningful studies on the physiology and biochemistry of
zoospore formation have been achieved with members of the Blastocladiales, in particular *Blastocladiella* and *Allomyces* (Cantino, 1966).
Members of the Blastocladiales grow in water environments and in soil,
and are characterized by the production of thick-walled, resistant
sporangia which usually have pitted walls. The zoospore is distinctive in
having a prominent nuclear cap situated near the centre of the spore.
The genus *Allomyces* has a complicated and varied life cycle, and
although many studies have been made on zoospore biogenesis it has
primarily been investigated with respect to hormonal control of
sexuality (Chapter 7).

    The genus *Blastocladiella*, and in particular *B. emersonii*, has become the most experimentally exploited fungus for studying the factors initiating and regulating the process of zoospore morphogenesis. The zoospore displays a remarkable totipotency in that, depending on the environmental conditions, it can show four distinctly different pheno- typic developmental patterns: thin-walled ordinary colourless sporan- gium (OC), thin-walled orange sporangium, thick-walled resistant sporangium (RS), or thin-walled late colourless sporangium (Fig. 6.10; 6.11). The zoospores are posteriorly uniflagellate and after a period of high motility they settle down, retract the flagellum and develop a uni- nucleate germ tube which forms a branched rhizoidal system. After an exponential phase of increase in dry weight, volume and other features the basal cell has enlarged several hundredfold. A cross-wall is laid down which delimits the thallus into a basal rhizoid-bearing cell and an apical multinucleate fertile cell. Within the apical cell rapid mitosis occurs and the cell protoplasm becomes partitioned into hundreds of uninucleate

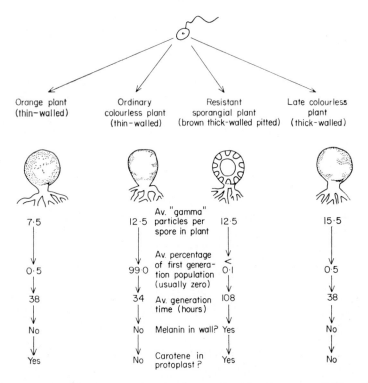

Fig. 6.10. The four developmental paths that can be taken by spores of *Blastocladiella emersonii*, and the gross parameters which distinguish them from one another. From Cantino (1961).

haploid zoospores which are released through pores formed by the dissolution of several papillae.

When *Blastocladiella emersonii* is cultured in enclosed surface plates over 99% of the zoospores will germinate and give rise to OC sporangia. When such a first-generation OC sporangium is mature, it proliferates by liberating a crop of first-generation zoospores from which a clone of second-generation cells or sporangia are produced. Should an OC cell

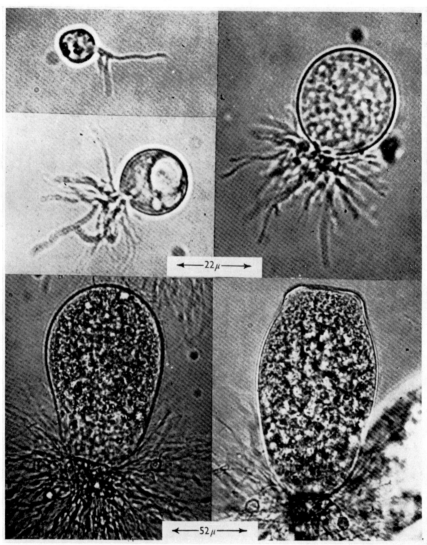

Fig. 6.11 (*a*)

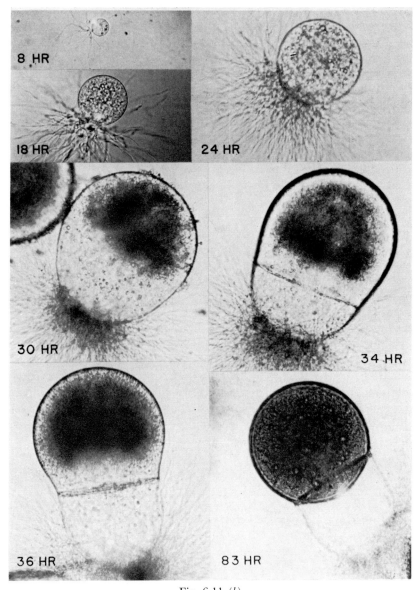

Fig. 6.11 (*b*)

Fig. 6.11. The two main developmental pathways in *Blastocladiella emersonii*. (*a*) The appearance of OC cells growing in synchronous, single-generation culture. From Goldstein and Cantino (1962). (*b*) The appearance of Rb cells growing in synchronous, single generation culture. From Lovett and Cantino (1960).

be removed from this clone and placed in a fresh medium the process will be repeated and in this way this cycle can continue *ad infinitum*. However, if the OC cells are left in their original cultural environment they will show this remarkable totipotency, developing in variable proportions into the four morphological forms. As proliferation continues the cell mass will gradually come to be composed of RS cells with a sparse distribution of other types. Thus the zoospore of *B. emersonii* from a fixed genome can express four different phenotypic patterns of development. This is indeed an exciting situation since it shows in a relatively simple organism how pattern formation could develop in multicellular organisms from similar stem cells.

The totipotency of the OC cells obviously represented a serious impediment to a fuller comprehension of zoospore biogenesis. However, the discovery by Cantino in the 1950s that the presence of bicarbonate could lead to almost 100% development of RS cells, whereas essentially all the zoospores developed along the OC pathway in the absence of bicarbonate provided a system in which the "trigger" reaction leading to one type of differentiation rather than to another could be readily analysed. This observation led to the development of submerged, synchronized, single generation cultures containing up to 10^9 individual cells and provided an elegant system for studying the relations between biochemical and morphological differentiation.

The original studies by Cantino and his associates suggested that the metabolism of the OC cells was predominantly homofermentative; a condition that may have been influenced by the poor aeration in the early experimental system. The potential of the tricarboxylic acid cycle in supplying energy was examined, and it was considered that although it was present it played only a minor role in this respect. Their studies further showed that when bicarbonate was added to the developing OC cells it caused a set of multiple enzymic lesions in the tricarboxylic acid cycle. However, they found that isocitrate deyhdrogenase specific for NADPH[2] remained functional and began to operate in reverse, mediating the reductive carboxylation of α-oxoglutarate to isocitrate. Bicarbonate also appeared to induce the formation of isocitrate lyase which cleaves isocitrate to glyoxylate and succinate. Isocitrate lyase represents part of an anaplerotic pathway found in most organisms. Finally, a constitutive glycine-alanine transaminase brings about the formation of glycine by the amination of glyoxylate. The glycine was then considered to be utilized in the extensive RNA synthesis that accompanied RS development. The bicarbonate trigger mechanism is shown in Fig. 6.12.

During the development of the RS sporangium a point is reached

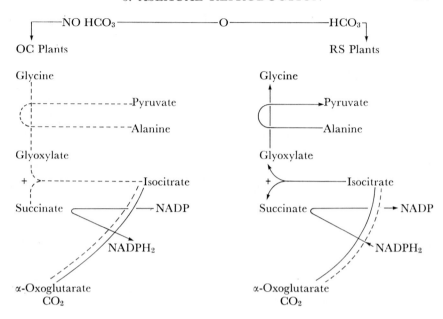

Solid lines indicate major enzyme activity.

Fig. 6.12. The bicarbonate trigger mechanism in *Blastocladiella emersonii*. Solid lines indicate major enzyme activity. From Cantino (1961).

beyond which the cell becomes irreversibly committed in its developmental pathway. This can be considered as the morphological point of no return. Up to this point, which is 43% of the generation time in the conditions used by Cantino, removal of bicarbonate from the medium caused reversion of the developmental pathway and a return to OC development. Beyond this point of no return, which coincided with a cessation of increase in cell size and the completion of cross-wall formation, the presence or absence of bicarbonate did not affect the ultimate nature of the cell. Many of the morphological and biochemical events that occur during RS development both before and after the point of no return have been documented and are illustrated in Fig. 6.13.

Do any of the features associated with an RS cell, during exponential cell development, revert to those more representative of OC development, if bicarbonate is removed before the point of no return? In fact, some but not all of the RS characteristics do revert, in particular the two key enzyme systems considered to be mainly responsible for RS formation. After the point of no return, removal of bicarbonate has no effect on the enzymes. Cantino (1966) has considered that no one single

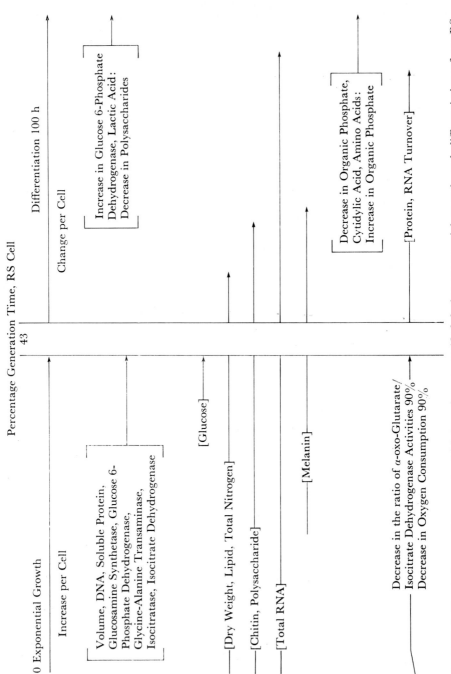

Fig. 6.13. A digest of some of the events which have been quantified during exponential growth and differentiation of an RS cell of *Blastocladiella emersonii*. From Cantino (1967).

factor can be responsible for controlling development but rather it is the cumulative result of many biochemical events each having its own point of no return dispersed over a much broader range of time.

Thus in summary the Cantino investigations would imply that bicarbonate, as a result of the reductive carboxylation of α-oxoglutarate, causes multiple enzymic lesions in the tricarboxylic acid cycle with the consequent accumulation of intermediates which can then be shunted to other reactions necessary for the genesis of RS cells. Furthermore, the bicarbonate effect was found to be related to critical periods in the development of both RS and OC development.

Further studies on this developmental system by Khuow and McCurdy (1969) have produced results which differ from predictions based on Cantino's hypothesis. They have shown that all tricarboxylic acid enzymes, except α-oxoglutarate dehydrogenase, can be detected in extracts of RS and OC cells and that these enzymes have a higher activity per organism in RS cells (Fig. 6.14). Since the RS cells are larger than the OC cells specific activity measurements did however show that most of the enzyme activities were lower in RS cells. Specific activities reached a maximum at 24 h except for NADP-dependent iso-citrate dehydrogenase which increased throughout the 36-h period of growth. Thus these results contradict the findings of the Cantino group and imply that the tricarboxylic acid cycle enzymes are present in RS cells during development and furthermore indicate that the bicarbonate effect does not lead to the loss or significant decrease of tricarboxylic acid cycle enzymes during the crucial period of induction. Indeed there is a striking similarity between the activity of the tricarboxylic acid cycle enzymes and many other characteristics of RS and OC cells during the exponential period of growth.

It would appear that the main difference between these studies is the fact that the Cantino group compared OC and RS cells on the basis of generation times which assumes that their respective maturation times reflect the activities of analogous stages of growth and development. In fact, exponential growth continues throughout the development of OC cells whereas in RS cells it precedes a prolonged period of development in the absence of growth. Khuow and McCurdy (1969) consider that meaningful comparisons can only be made during the exponential growth phases of RS and OC, and in this context there appears to be no development of multiple lesions in the tricarboxylic acid cycle of RS cells. Thus an explanation for the trigger mechanism of bicarbonate must yet be found.

Radioactive tracer studies using ^{14}C labelled bicarbonate showed that significant amounts of the label appeared in aspartate with lesser amounts in glycine and malate. Since aspartic acid was the first compound to be labelled, and it is formed rapidly from oxaloacetic acid, it has been proposed by Khuow and McCurdy that oxaloacetic acid may be the initial product of bicarbonate fixation and that phosphoenol-

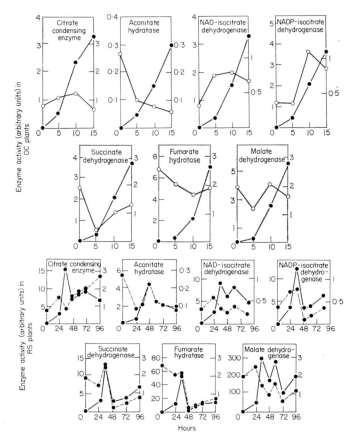

Fig. 6.14. Relationship between enzyme activity and development of *Blastocladiella emersonii* in OC and RS cells. In data for OC plants, ○ indicates units of enzyme activity per mg protein, and ● units per plant $\times 10^6$. In data for RS plants, ●——● indicates units per mg protein, and ●——● units per plant $\times 10^6$. From Khouw and McCurdy (1969).

pyruvate or pyruvate is the actual site of CO_2 fixation. The metabolism of bicarbonate would then be dependent upon a functional tricarboxylic acid cycle, and on the relatively high levels of the corresponding enzymes existing before the point of no return. After this point, a decrease in the activities of these enzymes and a decreased rate of carbon dioxide fixation may be causally related to morphogenesis. Evidently, the biochemical changes initiating and controlling the developmental pattern within *Blastocladiella emersonii* are not yet clearly understood.

Methods have been developed for growing and treating large populations of mycelia of *Achlya* so that the hyphal tips will differentiate into sporangia with considerable synchrony (Griffin and Breuker, 1969). Under the experimental conditions imposed for the differentiation of the sporangia, net RNA, DNA and protein synthesis ceased. However, throughout the period of differentiation there was continued incorporation of radioactive precursors into RNA. The concentration of actinomycin D used did not inhibit the normal growth and branching of the mycelia that occurred during differentiation. Clearly, DNA-dependent RNA synthesis was required for the complete differentiation of sporangia. Sucrose-gradient analysis of newly synthesized RNA showed that only the ribosomal and soluble fractions of RNA were labelled during vegetative growth. During differentiation of sporangia, ribosomal and soluble RNA fractions were also labelled, and, in addition, a heterodisperse fraction of labelled RNA which was heavier that ribosomal RNA appeared. The patterns of incorporation undoubtedly showed that synthesis of ribosomal RNA continued up to the time of spore discharge in *Achlya* which differs from the situation in *Blastocladiella emersonii* where RNA synthesis ceases during differentiation of the sporangia.

The short duration of zoospore formation and the ability to control development by careful manipulation of the environmental conditions makes such organisms excellent experimental systems. It is surprising that there have been no serious attempts with *Blastocladiella* and related fungi to study the biochemistry of membrane formation. Few organisms can possess such large concentrations of new membranes relative to other cellular constituents as occurs during zoospore formation. Several excellent ultrastructural studies have been carried out during zoospore formation (for ref. see Gay and Greenwood 1966).

Sporangiospores

Relatively few studies have been carried out on the biochemistry of sporangiospore formation. A study of enzyme localization using the nitro-blue tetrazolium staining technique has shown that the tricarboxylic acid cycle is active in young stages of differentiation of *Mucor hiemalis*. Glutamic acid dehydrogenase is also present in young sporangiophores.

The ultrastructural features of sporogenesis have been described for

Gilbertella persicaria (Bracker, 1968). The reorganization of the proto-plast into a multinucleate unit involves changes in cytoplasmic mem-branes. Cleavage of the protoplast is initiated by the fusion of vesicles to form an anastomosing tubular network. The membrane bound net-work becomes dilated and delineates a system of cleavage planes and the outlines of the new cells. As the process continues elements of the network converge to complete cleavage. The vesicles have stainable contents and are associated with the endoplasmic reticulum. Evidence from other sources confirms the involvement of vesicles in the delinea-tion of endogenous spores in many fungi.

There is a distinct contrast between sporangial cleavage and the for-mation of asci. During ascus formation a peripheral double membrane invaginates to delimit spore initials.

As yet it is not possible to trace the origin of fungal membranes during development. Zygomycete fungi as well as many others have no mor-phologically defined Golgi apparatus. Indeed, true Golgi apparatus occurs only in members of the Oömycetes (see Chapter 5). In *Pythium* the origin of the vesicles can be readily seen. What cell component then carries out the functions of the Golgi apparatus when none is evident? Does some other cell component take over the secretory role? Bracker considers that the ring-like cisternal profiles clearly visible in most sections of *Gilbertella* are the functional equivalent of the Golgi apparatus. The membrane bound cisternae are then capable of packaging material within a membrane for transfer to an extracellular environment. An intracellular component becomes an extracellular secretion during the formation of new cells. Thus again we see the important part played by vesicles during a morphological event.

Biogenesis of Conidia

Many filamentous fungi are characterized by the formation of thin-walled cells called conidia which can become easily separated from the parent mycelium. Such conidia can arise by the formation of terminal or lateral cells from special hyphae—the conidiophores. The conidio-phores may be morphologically very similar or very different from vegetative hyphae. At their most complex the conidiophores can be thick-walled, erect and perpendicular from the vegetative mycelium and be composed of many cells. Two types of conidium production have been extensively studied, viz. *Aspergillus* and *Neurospora*. The conidio-phore of *Aspergillus* spp. is undoubtedly among the most complex in the fungi whereas in *Neurospora* the conidiophore is relatively undifferenti-ated and is not unlike vegetative mycelium. In both types conidiogenesis occurs after the rapid phase of growth and the consequent exhaustion of the limiting nutrient.

In *Neurospora* the vegetative mycelium can, during the course of development, initiate two types of conidia: multinucleate macroconidia or uninucleate microconidia. Both macroconidia and microconidia of heterothallic species of *Neurospora* can function not only for asexual propagation but also in sexual reproduction where they can serve as the male nuclear donor in spermatization. The macroconidia

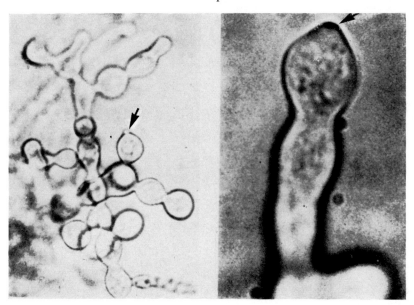

Fig. 6.15. Conidial development in *Neurospora crassa*. (a). "Dry" aerial proconidia chains of *N. crassa* from 2-days-old cultures on solid C-medium showing budding process on a narrow base (arrow). A first septum closing a narrow isthmus is visible on the right. ×850. (b). A proconidial chain of *N. crassa* developed on a 2-days-old liquid C-culture showing a wide budding base initiated as a papilla (arrow) ×1800. From Turian and Bianchi (1971).

may be technically called blastoarthrospores, i.e. blastospores basifugally budding on limited conidiophores and becoming secondarily disarticulated from the proconidial chain as arthrosporal elements (Fig. 6.15).

In *Aspergillus* the first morphological indication of asexual reproduction is the enlargement of certain cells in the mycelium. These cells, termed foot cells, produce a single erect, thick-walled stalk cell, the conidiophore, as a branch perpendicular to the long axis of the cell (Fig. 6.16). Conidiophores are generally unbranched and non-septate.

During elongation and growth the conidiophore tip is pointed but then enlarges into a bulbous head, the vesicle. As this now multinucleate vesicle develops a large number of fertile cells or sterigmata are produced over its surface either parallel and clustered in terminal groups or radially from the entire surface. The sterigmata may occur as one series

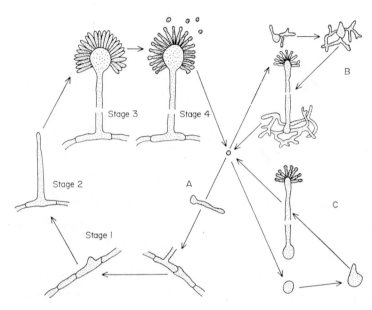

Fig. 6.16. Summary of induced morphogenetic sequences leading to conidiophore development in *Aspergillus niger* under submerged agitated conditions. A. Sequence of morphological changes in replacement fermenter culture. Stage 1, conidiophore initiation; stage 2, conidiophore elongation; stage 3, vesicle and phialide formation; stage 4, conidiospore production. B, C. Forms of microcycle conidiation. In B, a branched mycelial system and a mature conidiophore are produced from an enlarged conidium. Treatment consists of incubation at 41° for 15 h followed by 30° for 12 h. In C, a mature conidiophore is produced from an enlarged conidium in the complete absence of vegetative development. Treatment consists of incubation at 44° for 48 h followed by 30° for 15 hours. From Smith and Anderson (1973).

or as a double series with each primary sterigma bearing a cluster of several secondary sterigmata or phialides at the apex. From these phialides the conidia are developed and in still air conditions long chains of conidia will develop. Phialides are typically bottle-shaped and the conidia develop as endogenous phialospores, formed in basipetal succession inside the phialide by transverse septation, and are extruded at the tip. The vesicle generally supplies the initial complement of one or more nuclei to the phialide. Asexual formation of conidia can occur on surface culture or in submerged culture where conidiophores are reduced in length but are otherwise normal and grow out from mycelial pellets or from units of filamentous mycelium. A novel microcycle conidiation has been obtained with several *Aspergillus* species in which the conidiophores develop directly from an enlarged conidium without an intervening mycelial stage.

One of the main problems in assessing the many studies that have been carried out on conidiogenesis in filamentous fungi is the wide range of experimental systems that have been used. These conditions range from solid plate surface culture through static liquid to submerged liquid agitated cultures with highly refined monitoring controls. Because of these variable growth conditions it is difficult to formulate any clear-cut theories on the biogenesis of conidia. However, it is clear that if meaningful interpretations are to be made on this type of differentiation, then the experimental system must impose rigorous control over the developmental pattern. Only then will it be possible to achieve a good understanding of the factors that control conidium formation in filamentous fungi.

How essential is the complex conidiophore? Apparently it can be by-passed to a limited degree with modified vegetative hyphae cutting off conidi. However, conidium production by this method cannot compare on a quantitative basis with that of the large, complex conidiophores. Such conidiophores probably only occur under certain environmental conditions and are not obligatory to conidium production.

CULTURAL CONDITIONS FOR CONTROLLED CONIDIUM FORMATION

In the studies which have been carried out on conidiation in filamentous fungi, increasing emphasis has been placed on the development and refinement of culture techniques. In many instances improvements in understanding of differentiation processes have rapidly followed the introduction of culture methods which have increased the measure of control over morphogenesis. The use of defined media for growth, sophisticated culture vessels and the application of continuous flow culture techniques have allowed a precise analysis of the environmental factors which affect morphogenesis.

Synchrony as defined by conventional use is the simultaneous division or doubling of a cell population. Synchronous cellular division can be obtained with bacteria, protozoa and some plant and animal cells in tissue culture. Fungi which grow as coenocytic multinucleate hyphal tubes do not lend themselves to the development of synchrony in cell divisions. However, besides cell division, there are other events such as morphological differentiation which occur during development and which can now be synchronized. A better understanding of how development and differentiation occur in filamentous organisms can be achieved in such systems where changes in morphology are synchronous for a given culture.

Conidiation in surface culture: Growth on the surface of a suitable medium is the traditional cultural method for filamentous fungi. Inoculation of the fungus at a point on a solid nutrient medium results in the formation of a circular colony which is characterized by zones of differing morphology and metabolic activities. For *Aspergillus niger* the onset of conidiation in these colonies begins some distance behind the extending margin and there occurs a range of developmental states from immature conidiophores in the outer regions of the colony to aged conidiophores in the centre (Fig. 6.17). The zonation effects which are characteristic of colony development from a point inoculum are eliminated by spraying the inoculum evenly over the surface of the medium. This method of inoculation has been effectively used to produce large quantities of mycelium and conidiophores of a more uniform age for studies on biochemical changes during differentiation. In these studies the fungus is grown on a cellophane membrane placed on the surface of the solidified medium to simplify the harvesting of intact hyphal mats. A technique involving lyophilization of the mat followed by brushing has been used to selectively harvest only the aerial conidiophores for biochemical studies. Conidial inoculum may also be mixed with a semi-solid medium and then deposited as a shallow layer on a base of solidified medium. The mycelial mat so formed can be readily peeled away from the underlying agar. Conditions can be controlled in such cultures by covering the vegetative mycelium with a layer of cellophane. When the cellophane is removed there is rapid and uniform development of conidiophores over the surface of the mycelial mat.

The fungus *Trichoderma viride* requires a photo-induction of vegetative mycelium to produce conidia. In this system a mycelial inoculum is plated in the centre of a Petri-dish containing a suitable medium and placed in the dark where a radially growing colourless colony develops. When sufficient mycelial growth is present on the plate the colony is illuminated for less than one minute. A dark green ring of conidiophores with conidia then emerges at what was the growth perimeter at the time of illumination, while vegetative growth continues radially. The morphological changes can be seen microscopically approximately 3 h after photoinduction and the mature conidia appear one day later.

A vertical section of any of these surface colonies will show a lower dense mycelial layer which penetrates to a greater or lesser extent into the substratum, and an upper layer composed of loosely interwoven aerial hyphae. The conidiophores grow upwards away from the medium, and final differentiation of the conidial heads takes place in an aerial

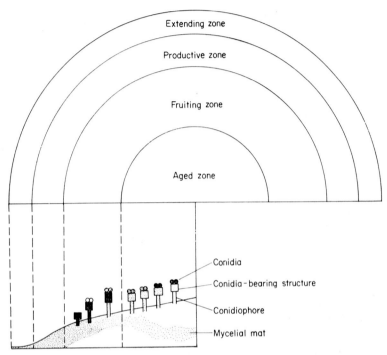

Fig. 6.17. Schematic diagram showing the four concentric differentiation zones in the surface colony of *Aspergillus niger*. The extent of conidiophore maturation in each of the zones is shown by the localization of basophilic substances (represented as dots of varying density). These substances are highly concentrated in the mycelium of the extending zone and in the early stages of conidiophore development. As the conidiophores mature the basophilic substances progressively move from the conidiophores into the conidia. From Yanagita and Kogane (1962).

environment. Varying the environmental conditions can profoundly alter the developmental sequences (Smith and Anderson, 1973).

While environmental factors can modify the morphology of the conidial apparatus, the correct genetic potential is a prerequisite for normal conidiation. The difficulties of studying the genetic control of morphogenesis are well illustrated by the studies on the genetics of conidiation in *Aspergillus nidulans* (Martinelli and Clutterbuck, 1971). Using a formula developed to estimate the frequency of genes controlling sporulation in *Bacillus subtilis* it was estimated that between 300 and

800 genes were concerned with conidiation. Conidial mutations were observed between 6 and 13 times as frequently as auxotrophic mutations. However, in 85% of the mutants isolated the effect of the mutation was not restricted to conidiation. Such genes are pleiotropic affecting both growth and conidiation.

Many of the mutants isolated caused a reduction in the number of conidia produced rather than being totally asporogenous. Whereas it is relatively easy to locate the point at which morphogenesis is blocked by an asporogenous mutant, it is not so easy when some mature conidia are produced. Five groups of asporogenous mutants have been recognized depending upon the stage of development which is affected— (1) foot cell development; (2) vesicle formation; (3) sterigmata formation; (4) conidial bud formation; (5) conidial maturation. Conidial maturation mutants can be further subdivided into (a) normal appearance but conidia do not grow and (b) abnormal appearance.

The biochemical lesions associated with these mutations have not been recognized. All the mutants which were not completely asporogenous showed abnormal development from an early stage not later than the period of foot cell formation.

A further category of mutants is easily recognizable, namely those affecting conidium colour. Yellow, white, pale, yellow-green and chartreuse spores have been observed. The chemistry of the pigments involved has not been worked out although it has been demonstrated that yellow spored strains lack the enzyme p-diphenol oxidase (laccase). This enzyme is normally found in the wild type only during conidiation (Clutterbuck, 1972).

Several types of morphological mutants are known in *Neurospora*. Some have lost the ability to form conidia, e.g. aconidial fluffy strains, while others can only form one type of conidium.

Conidiation in submerged culture. Although surface culture is adequate for studies on the environmental and genetic factors affecting conidiophore development it has severe limitations for most nutritional and biochemical studies on conidiation. The physical nature of the mycelial mat creates an unavoidable degree of physiological variation within the cultures. Submerged agitated cultures will give more homogeneous conditions particularly if growth is in the filamentous rather than the pellet form. However, in submerged conditions most filamentous fungi remain entirely vegetative and this is consistent with the finding that most differentiated structures are characteristic of aerial mycelium in the surface colonies. The nature of the inhibition in submerged conditions may be complex and probably involves one or a combination of several factors such as reduction in oxygen tension, changes in the physical nature of the hyphal walls associated with submergence and direct contact of the conidiogenous cells with inhibitory factors in the medium.

Conidiation of *Aspergillus* species has been achieved in submerged shaken culture by manipulation of the culture medium. In submerged liquid culture it is possible to obtain more precise information on the relationship between the nutritional status of the medium and the onset of conidiation. In all such studies it has been shown that conidiation occurs at the end of the rapid phase of growth after the exhaustion of the limiting nutrient (Fig. 6.18). In this context the extensive studies on *A. niger* by Galbraith and Smith (1969) have provided the most detailed information on the induction of submerged conidiation in *Aspergillus* fungi. When glucose is the limiting nutrient, conidial induction is affected by the type of nitrogen source. Conidiation does not occur in the presence of excess ammonium ions in spite of glucose exhaustion although nitrate has no such inhibitory effect. Acetate, pyruvate, tricarboxylic acid cycle intermediates, glyoxylate and most amino acids are able to overcome ammonium ion inhibition of conidiation.

By altering the nitrogen source of a simple liquid synthetic medium

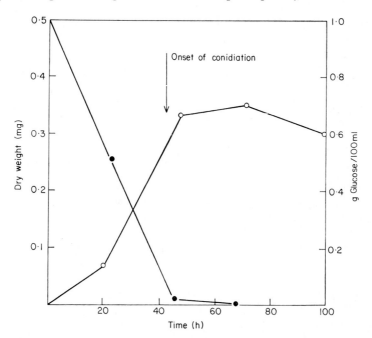

Fig. 6.18. Relationship between nutritional status of medium and the onset of conidiation; ●——●, carbohydrate utilization; ○——○, increase in dry weight.

enriched with citrate it has been possible to regulate the interconversion between the mycelial state and the conidiogenic state in *Neurospora*. When nitrate is the sole nitrogen source in the medium excellent conidiogenesis can be obtained in the upper layer of the mycelial mat (C cultures). Substitution of an ammonium salt for the nitrate causes the development to follow a purely vegetative direction (M cultures). In stationary cultures in M medium the presence of ammonia appears to promote certain physico-chemical changes in the mycelium which prevent the hyphae from breaking through the surface film of the liquid. In contrast, when M medium is aerated and agitated the mycelium can be induced to form conidia.

On the basis of these findings it is possible to compare the biochemical events in differentiating and non-differentiating mycelia by utilizing culture media which differ intially in only one factor.

An aerial stimulation for induction of conidiogenesis has also been proposed for *Penicillium* (Morton, 1961). He postulated that a surface-active material, possibly protein, rapidly forms when the mycelium becomes aerial and that production of such a substance is reversed by re-submergence. He considers that the essential difference between the aerial and submerged mycelium is the sudden formation of an air/water interface, which possibly causes changes in the orientation of the polar molecules such as the unfolding of protein which would result in meta-bolic change. In *Neurospora* there is also a change in the chemical nature of the surface of the conidiogenic hyphae on emergence, i.e. an accumu-lation of lipoproteins which give the aerial hyphae their hydrophobic character.

The above methods provide control over the induction of conidiation at the end of the growth phase in submerged shake culture but do not influence further maturation of the conidiophores. A much greater control over both the induction of conidiophores and their subsequent development has now been obtained with *Aspergillus niger* by using a fermenter culture system involving sequential medium replacement (Anderson and Smith, 1971a). By means of this technique cultural con-ditions which selectively favour each stage of conidiophore development can be introduced. On this basis four major developmental stages have been characterized for conidiophore maturation in *A. niger*. The first morphological stage of conidiation (stage 1), foot cell formation, is in-duced by growth in a medium with nitrogen as the limiting nutrient (LN medium). Foot cells cannot be induced in a high nitrogen medium in which both the nitrogen and carbon sources become exhausted simultaneously or in LN medium when the oxygen supply to the culture is reduced to a level which first maintained vegetative growth

(LNO medium). The second stage of development (stage 2), conidio-phore elongation, occurs in LN medium after the exhaustion of exo-genous nitrogen although continued metabolism of the carbon source is required (Fig. 6.16). If the mycelium is allowed to remain in this medium no further development takes place. Replacement of the culture to a new medium containing a nitrogen source and a tricarbo-xylic acid cycle intermediate as a carbon source effectively induces the third stage (stage 3), vesicle and phialide formation. Although the conidial apparatus appears to be fully developed at this stage no conidiospore production occurs from the phialides. This final fourth stage (stage 4) of development is most effectively induced by transfer to a medium in which glucose is the carbon source and nitrate the nitrogen source. If the culture is not replaced after stage 3 extensive prolifera-tions of the phialides occur. This control of the successive structural changes by media replacement results in the synchronous maturation of the conidiophores.

Stine and Clark (1967) have been able to achieve some degree of morphological and probably biochemical synchrony during *Neurospora* conidiation. Conidia are germinated in a minimal medium, and after 48 h incubation the mycelial growth is collected, washed and placed in moist Petri dishes. Conidiophores synchronously appear over the mycelial mat at 51 h and elongate to maturity to 56 h. At 56 h the conidiophores synchronously differentiate their first conidia and con-tinue to produce conidia for the next 8 h. At the time of the initial burst of conidial differentiation each conidiophore has a single conidium. With time, each conidiophore becomes a chain of individual conidia and this accounts for the increase in numbers of conidia produced with time. The production of conidiophores and conidia in synchrony and in large quantities allows sufficient experimental material from which to obtain meaningful results on the biochemical changes associated with differentiation. However, it must be noted that much of the biomass is vegetative and may well dilute some of the less dramatic changes associated with conidiogenesis.

Growth in submerged agitated batch culture clearly allows a much more precise examination of the effects of medium composition on differentiation than the static surface growth method. Nevertheless, the interpretation of differentiation responses to particular cultural condi-tions is complicated in the batch system by the transient nature of the environmental conditions during growth. For example, it is not possible in a batch culture system to determine whether conidiation results from nutrient limitation or from the limitation of growth rate imposed by this condition.

Chemostat culture will permit a study of microbial populations at various growth rates and under various metabolic steady states. In some studies using glucose-limited chemostat culture free conidia have been observed in the medium although differentiated conidiophores were not produced. Studies of the morphology of *Aspergillus niger* in carbon limited and nitrogen limited culture have recently been made (Ng., Smith and McIntosh, 1972). With this fungus, maximum conidiation occurs at growth rates between zero and the critical growth rate above which steady state vegetative growth prevails. When citrate is supplied, the carbon source limitation of this nutrient gives a much higher degree of conidiation than is obtained under glucose limitation. Under ammonium limitation with citrate as the carbon source there is no conidiation. When nitrate is used as the limiting nitrogen source conidiophore initiation but not maturation occurs. These experiments demonstrate that conidiation in *A. niger* can be controlled by growth rate only in particular media. Conidiation in this fungus appears to be determined by an interaction between growth rate and the nature of the carbon and nitrogen sources in the culture medium.

In citrate limited culture the number of free conidia in the culture medium decreases as the dilution rate increases (Table 6.1). This decrease in number of conidia could reflect either a decrease in conidium production or an increase in the flow rate so that more conidia are being washed out, or to a combination of these factors. However, the number of free conidia per gram dry weight of mycelium has a higher value at low dilution rates than at higher dilution rates, implying that conidiation intensity varies inversely with dilution rate.

An interesting feature of the chemostat culture of *Aspergillus niger* is the occurrence of a considerable reduction in the complexity of the conidiophore. Previous observations have shown that the conidiophore of *A. niger* produced in submerged culture under a variety of conditions is essentially similar, although smaller than, the normal subaerial

TABLE 6.1. Numbers of free conidia of *Aspergillus niger* in citrate-limiting chemostat culture. From Ng, Smith and McIntosh (1972).

Dilution rate (D)	Conidia/ml medium	Mycelium g dry wt/ml medium	Conidia g dry wt
0·012	3.5×10^5	$0·99 \times 10^{-3}$	$3·53 \times 10^8$
0·020	$1·7 \times 10^5$	$1·32 \times 10^{-3}$	1.29×10^8
0·029*	$0·42 \times 10^5$	$1·42 \times 10^{-3}$	$0·296 \times 10^8$

* Above D = 0.029 the number of free conidia was too low for accurate counting.

structure. In chemostat culture conidiophores were characterized by possessing small vesicles with few phialides and occasionally conidia were observed to develop from modified hyphal tips. Somewhat similar results have also been shown for *Penicillium* in chemostat culture. These reductions in conidiophore complexity may represent only a partial switch on of the conidiation mechanism. It does indicate that under certain conditions the morphological and biochemical events of conidiophore development which precede conidiospore formation can be by-passed. Morphological mutants of *Aspergillus nidulans* have been shown to produce normal conidia on abnormal simplified conidiophores.

MICROCYCLE CONIDIATION

The major limitation with these methods of cultivation is that conidiation is preceded by a period of vegetative filamentous growth which inevitably creates a heterogeneous cell population. A unique microcycle conidiation technique has recently been developed which, by eliminating the normal hyphal vegetative growth phase of the fungus, promises a novel approach to studies on the mechanism of conidiation in the aspergilli.

When conidia of *Aspergillus niger* are cultured under submerged conditions at elevated temperatures (38°–44°C) in a basal medium supplemented with glutamate they undergo striking morphological changes (Anderson and Smith, 1972). Whereas all conidia produce germ tubes at 30°, at temperatures from 38° to 43° the proportion of conidia which produce germ tubes progressively decreases and at 44° germ tube formation is completely inhibited. However, at this temperature swelling of the conidia continues to occur over a prolonged period to produce large spherical cells (20 μm mean diameter) (Figs. 4.2, 4.3 and Table 6.2). These cultural conditions also induce morphogenetic changes in the conidia which lead to the direct outgrowth of conidiophores (Anderson and Smith, 1971b). A prolonged (48 h) period of incubation at this temperature followed by incubation at 30° results in the direct outgrowth of a conidiophore from the enlarged conidium in the complete absence of normal vegetative growth (Figs. 6.19 and 6.20). Initially a single conidiophore develops and this is then frequently followed by a second and occasionally up to five are produced by the enlarged conidium. The conidiophores so produced are similar to, but smaller than, the normal subaerial conidiophores and produce viable conidia (Fig. 6.21).

The complete loss of the ability of the conidia to produce vegetative growth in a complete growth medium at 30° after a prolonged period of

TABLE 6.2. Effect of spore-population density at different temperatures on the formation of SG spores, the degree of spherical growth, and the degree of germ-tube formation from spores of *Aspergillus niger* after 15 h cultivation. From Anderson and Smith (1972).

Spores per milliliter medium, $\times 10^6$	No. of SG spores, % Temperature			Size of SG spores μ Temperature			No. of SG spores with germ tubes, % Temperature		
	30°	41°	44°	30°	41°	44°	30°	41°	44°
4	51	32	30	10·4	14·0	9·2	97	37	0
3	73	76	64	9·5	15·4	9·8	96	30	0
2	98	90	75	9·5	15·0	10·2	99	52	0
1	97	95	93	9·0	13·8	9·7	96	75	0
0·5	99	98	95	7·8	13·0	9·3	100	91	0
0·1	100	97	94	7·1	12·1	7·3	97	99	0

enlargement at 44°, whilst retaining the ability to form a complex reproductive structure and viable conidia, is extremely puzzling. However, studies on this phenomenon should provide valuable information relating to the control of both the vegetative and asexual reproductive phases of this fungus. Also, as a culture technique for studies on conidiation it should simplify the studies of regulatory mechanisms preceding conidiation and of the processes accompanying conidiophore development and conidia production.

BIOCHEMISTRY OF CONIDIATION

Filamentous, multicellular fungi are inherently difficult systems from which to obtain meaningful results on specific biochemical changes associated with the process of differentiation. As with other microbial systems, most physiological and biochemical studies have been concerned with the analysis of crude homogenates of the entire culture. Such methods may closely indicate the true changes occurring at specific phases of development in synchronously growing unicellular cultures. However, with mycelial cultures, and in particular because of the phenomenon of apical growth, there exists within each mycelium and indeed each hypha a spatial distribution of differing biochemical activities. As a result, particular aspects of fungal differentiation, e.g. conidiation, may not necessarily involve the entire thallus, and since normally the vegetative cells far outnumber the cells actively involved in conidiation important, specific and highly characteristic biochemical changes may be masked by the vegetative physiology. The problems

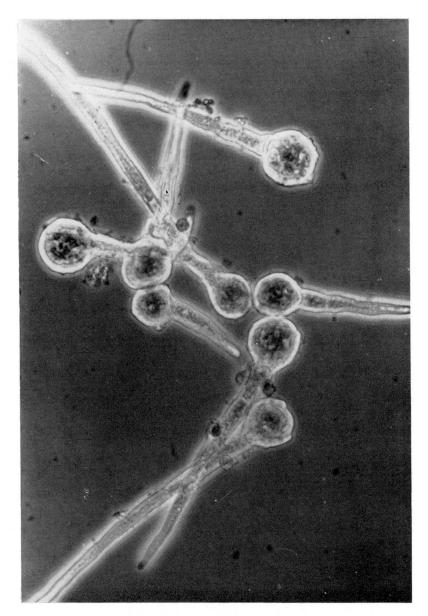

Fig. 6.19. Elongated conidiophore stalks emerging from enlarged conidia, 9 h after transfer to 30°. From Anderson and Smith (1971).

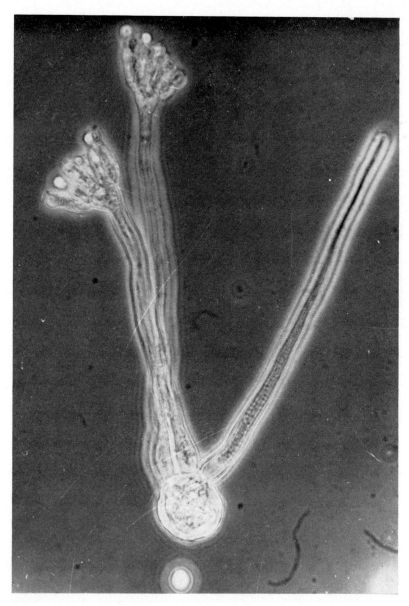

Fig. 6.20. Multiple conidiophore production (two mature and one immature) from an enlarged conidium, 18 h after transfer to 30°. From Anderson and Smith (1971).

inherent in the analytical methods are further multiplied in static mycelial cultures where variation of the physiological and biochemical status of the mycelium is increased by the additional variants of oxygen tension and nutrient concentration. Submerged batch or continuous cultivation reduces these complications only if the mycelium grows in a filamentous form rather than in pellet form, the more usual condition of growth in submerged cultures. With pellet growth nutrients and oxygen are only readily available to the peripheral hyphae and autolysis rapidly

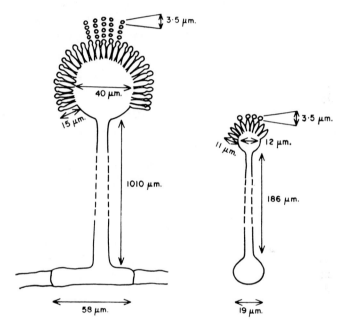

Fig. 6.21. Comparison of the typical subaerial conidiophore of *Aspergillus niger* produced from conidiating mycelium and from microcycle conidiation. From Anderson and Smith (1971).

occurs at the centre while dense growth continues at the edge. Although submerged growth is more desirable for biochemical studies of conidiation it is often beset with the problem that conidiation is generally suppressed in submerged culture even in those species which conidiate freely in static surface culture. However, it has clearly been shown that this difficulty can be overcome by careful manipulation of the cultural conditions.

Numerous studies have dealt specifically with quantitative and qualitative changes in enzyme levels during differentiation of several fungi and correlations between increasing levels of specific enzymes and the changing requirements of differentiation have also been described. *Carbon metabolism.* Oxidative metabolism appears to be essential to conidiogenesis in *Neurospora* and environmental conditions which favour glycolysis inhibit conidial development (Turian, 1969). Thus a four-day vegetative mycelium produces 80% more ethanol than a corresponding conidial mycelium; the addition of *p*-chloromercuribenzoate almost eliminates alcohol production and causes the M culture to conidiate. Inhibition of the tricarboxylic acid cycle with fluoroacetate favours glycolysis and C mycelium reverts to M mycelium.

Differences in relative enzyme activities in C and M cultures are as expected; thus alcohol dehydrogenase and carboxylase activities are greater in M mycelia. The occurrence of high glycolytic activity in filtrates M mycelium is confirmed by manipulation of the environment in various ways, such as by the induction of conidiation in an amino acid/ammonium medium by increased oxygen tension or by the addition of glycine to an ammonium medium. Therefore, conidiation is considered to be a morphogenetic expression of the Pasteur effect. Thus conidiation must be regulated by the balance between the oxidative and glycolytic pathways, probably at the point of pyruvate. The relative concentrations of reduced and oxidized NAD may also play a regulatory role. An active hexosemonophosphate pathway which appears during conidiogenesis would be expected to decrease the activity of the Embden-Meyerhof-Parnas pathway through competition for glucose-6-phosphate. In M hyphae, alcohol dehydrogenase has been cytochemically detected by oxidative assay and demonstrates a dense, uniform distribution of activity except at hyphal tips. In the conidiating hyphae, alcohol dehydrogenase becomes less dense in distribution especially in the budding apices. Cytochrome oxidase activity, localized in the mitochondria, is confined to the subapical zone of vegetative hyphae while at the initiation of conidiation it becomes dispersed throughout the proconidial buds.

During induction of conidiogenesis in *Neurospora crassa* by acetate, enzymic analyses demonstrate a much more active isocitrate lyase than in mycelium conidiating in sucrose medium. The increased conidiogenesis that occurs at 37°C compared with 25°C is also mirrored by a higher isocitrate lyase activity. Bicarbonate addition to the 25° culture also increases isocitrate lyase levels. Thus three different environmental conditions that induce increased conidiation also result in higher isocitrate lyase activity. The glyoxylate formed by isocitrate lyase is transaminated with alanine to form glycine.

However, it may well be that a source of glyoxylate rather than an active glyoxylate cycle is vital for conidial development since conidia are formed at 25° under conditions of relatively low isocitrate lyase activity. In this case, the glyoxylate could be formed by the splitting of pentose produced by the pentose phosphate pathway. The nitrogen source in these experiments is nitrate and its assimilation through nitrate reductase could provide a mechanism of NADP regeneration essential for the continued function of the pentose phosphate pathway. Addition of ammonium salts to the medium inhibits conidiation. NH_3 is preferentially utilized and probably uncouples nitrate reductase from the pentose phosphate pathway and prevents conidiation.

Induction of oxidative metabolism is essential for the expression of the conidiation potential; for example it can be induced by nitrate which has the effect of re-oxidizing $NADPH_2$ which can be coupled with glucose-6-phosphate dehydrogenase. A flavin type of metabolism coupled through $NADPH_2$-NADP regeneration to the direct oxidation of sugars via the hexose monophosphate pathway predominates during conidial differentiation. Several enzymes, including succinic dehydrogenase, NAD nucleotidase and NAD-dependent glutamate dehydrogenase, also show increased activites during conidiogenesis. Table 6.3 and Fig. 6.22 summarize the differences between purely vegetative mycelium and conidial cultures investigated for a number of biochemical characters (Turian, 1969).

Somewhat similar observations have been made with *Aspergillus* spp. $NADPH_2$-dependent isocitrate dehydrogenase and isocitrate lyase show much higher specific activities at the period preceding conidiophore development in *A. niger* than during vegetative growth of mycelium of the same physiological age. These enzymes are also active during conidiophore maturation in replacement culture (Fig. 6.23) and in continuous culture. Malate dehydrogenase, aconitase, NADP-dependent isocitrate dehydrogenase and malate synthetase are relatively similar in pre-conidiating and vegetative mycelium and do not appear to be highly activated during conidiation in replacement culture and continuous culture. In flask culture studies glycine-alanine transferase is detectable quantitatively only in pre-conidiating mycelium. It can be somewhat tentatively concluded that in *A. niger* the tricarboxylic acid cycle is only ticking over slowly during the rapid vegetative phase of growth due to catabolite repression, and since α-oxoglutarate dehydrogenase can never be detected it is possible that the tricarboxylic acid cycle serves purely a synthetic function. A major distinction between conidiating and non-conidiating mycelia in the systems so far examined is the synthesis of glyoxylate and possibly glycine through NADP-dependent isocitrate dehydrogenase and isocitrate lyase. The importance of glycine in RNA synthesis cannot be underestimated and the in-

TABLE 6.3. Biochemical analysis and physiological differentiation of mycelial and conidial cultures of *Neurospora crassa*. From Turian (1969)*.

Criterion	Mycelial cultures	Conidial cultures
	Specific activity	
Enzymes:		
Glucose 6-phosphate dehydrogenase	760	605†
NADP nucleotidase	309	12,618
NADPH₂-cytochrome c reductase	205	743
Succinate-cytochrome c reductase‡	648	771
Succinate dehydrogenase‡	92	109
Cytochrome oxidase	8,720	8,070
Isocitrate lyase	45	95
Pyruvate carboxylase	14	7
Ethanol dehydrogenase	1,659	125
	Specific production	
Chemical Composition:		
Ethanol§ (mg./g. dry wt.)	896	118
Acetaldehyde (mg./g. dry wt.)	2.1	0·3
Carotenoids (μg/% dry wt.)	290	12,000
Physiological activity:		
Q_{O_2}	12·8	18·9
CO_2/O_2	4·2	1·2

Age of cultures 3–4 days; Enzymes—specific activity; chemical composition = specific production.
*3–4 day-old cultures were examined.
† Value was underestimated by rapid destruction of NADP (see value for NADP nucleotidase).
‡ In isolated mitochondria instead of cell-free extracts.
§ Cultures were 4 days old.

creased activities of the enzymes which ultimately lead to increased glycine production may be intricately involved in regulating the RNA synthesis concerned with differentiation.

Recently there has been an extensive examination of carbon catabolism during differentiation of *Aspergillus* fungi. In these experiments, *in vitro* enzyme determinations were coupled with the radiorespirometric analysis of glucose metabolism *in vivo* in order to give a more reliable estimation of the *in vivo* changes occurring in glucose catabolism. It was found using the replacement fermenter technique that during conidiophore development the pentose phosphate pathway enzymes were higher in activity than the Embden-Meyerhof-Parnas pathway enzymes and this when taken together with the radiorespirometric analysis strongly implies that the direct oxidation of glucose

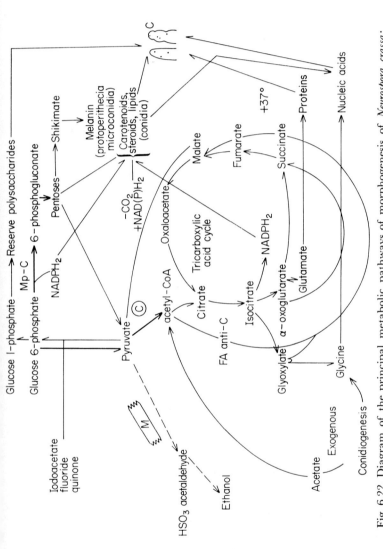

Fig. 6.22. Diagram of the principal metabolic pathways of morphogenesis of *Neurospora crassa*: carotenoid pigmented conidia, differentiation stimulated by extracellular acetate or glycine, increase in temperature (37°), or suppression of glycolysis. FA = fluoroacetate, an inhibitor of conidiation. M, Mp and C indicate predominant pathways associated with undifferentiated mycelium (M), with fertile mycelium producing protoperithecial asci (Mp) and microconidia, and with the differentiation of macroconidia (C). Note the double role in conidiation (C) of gluconeogenesis from acetate, and the reducing power of NADPH₂. (favourable to lipid synthesis). From Turian (1969).

through the pentose phosphate pathway may be of importance during conidiophore development. One of the main functions of the pentose phosphate pathway in cellular metabolism is to produce $NADPH_2$ essential for reductive biosynthesis.

Undoubtedly one of the main enzymic changes associated with conidiogenesis in filamentous fungi is the stimulation of the pentose phosphate pathway for carbon catabolism. The high biosynthetic demands of the process obviously cannot be met from normal vegetative metabolism, and this necessitates a major change in carbon catabolism. How are such changes in enzyme pathways brought about? In this case the enzymic pathways are already present and functional, and the onset

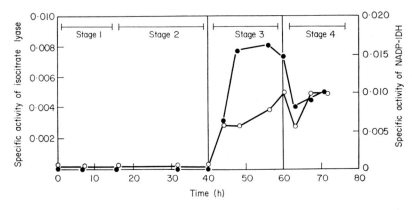

Fig. 6.23. Activities of NADP-isocitrate dehydrogenase and isocitrate lyase of *Aspergillus niger* during growth and morphogenesis in replacement fermenter culture. ●————●, Isocitrate lyase activity; ○————○, NADP-isocitrate dehydrogenase activity.

of conidiation leads to concomitant partial repression of one pathway and stimulation of the other. If it is considered that there is a causal relationship between the activity of these pathways and conidiation then how are these pathways regulated? Perhaps control of pathway activity is by way of critical levels of essential intermediates. Studies with *Aspergillus niger* during differentiation in several cultural conditions have shown that major changes can occur in the levels of certain intermediates of glycolysis. The fact that there was little apparent correlation between intermediate levels in the two systems would substantiate the results of Wright with *Dictyostelium discoideum* (Chapter 2) that the intracellular concentrations of certain metabolites essential to differentiation can vary from one study to another and yet normal morphogenesis can occur. This would imply that critical changes in the concentration of metabolites essential to differentiation can be met and balanced by compensatory mechanisms.

Further studies on a wider range of cellular intermediates may well be of value in determining the mechanism of initiation of conidiogenesis. *Esterase activity.* Studies with lipolytic esterases have established a causal relationship between a biochemical event and a morphological character in a fungus (Lloyd *et al.*, 1972). In flask cultures of *Aspergillus niger* esterases are always present in mycelial extracts during conidiation irrespective of the mode of induction but cannot be detected by electrophoresis in the vegetative mycelium of cultures which would ultimately conidiate or in mycelium of sterile cultures. In replacement

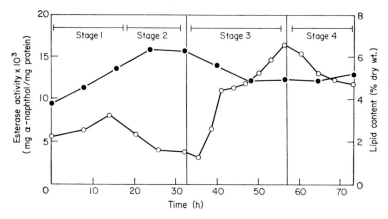

Fig. 6.24. Comparison of esterase activity and lipid content of Aspergillus niger during growth and morphogenesis in replacement fermenter culture. O————O, esterase activity of the organism; ●————●, lipid content of the organism. Lloyd *et al.* (1972).

fermenter techniques quantitative esterase determinations demonstrate a low basal level of esterase activity during vegetative development (Fig. 6.24). Esterase activity increases greatly immediately prior to vesicle and phialide formation and persists in these structures once formed. The increase in esterase activity during conidiation has also been shown cytochemically to occur in the conidiophore tip prior to the formation of the vesicle and phialides and in the latter structures after their formation. Such studies must imply that lipids function as a source of carbon and energy during conidiation. Esterase production is also associated with conidiation induced in continuous culture and in microcycle conidiation. That esterase activity plays an important role in conidiophore differentiation must now be accepted since the esterases are always present at conidiation irrespective of the method of induction.

Nucleic acid studies. Very few studies have been made to relate nuclear expression with conidiogenesis. This is undoubtedly due in part to the poverty of investigations on basic DNA/RNA studies in filamentous fungi.

Recently there have been several interesting studies with *Trichoderma viride*, a fungus which requires a light stimulation for conidiation. Because of this light dependency it is possible to precisely time the onset of conidiation. When 8-azaguanine or 5-fluorouracil are applied within a certain time range relative to light induction and removed thereafter, there is complete suppression of growth. The inhibition by 5-fluorouracil can be overcome by the addition of uracil. Chromatographic studies of RNA species suggest that the continuous synthesis of RNA during critical periods is a prerequisite for photo-induced conidiation. This may mean that new species of RNA are being transcribed at these periods.

The technique of DNA/RNA hybridization has been used extensively in bacterial studies where it has been possible to demonstrate that transcription of different RNA species occurs during specific growth phases. When this technique was used with differentiating *Trichoderma viride* it did reveal differences in RNA species during early vegetative growth but did not detect changes in RNA transcription directly following photo-induction (Stavy, Galum and Gressel, 1972). This lack of detection of photo-induced transcription may be explained by one or more of the following reasons: the transition is not regulated by transcriptional control but perhaps by translational control; new transcription does take place but only a very small fraction of the genome is transcribed and is not detected by the hybridization technique; the change in RNA transcription is substantial but restricted to the relatively few aerial hyphae which become conidiophores and is masked by the massive vegetative contribution.

Undoubtedly, the photo-induced conidiation demonstrated in *Trichoderma viride* makes this a valuable model for differentiation in the fungi. The rapid and exact timing of the light-mediated induction has immense experimental potential and it is surprising that this system has not yet been widely appreciated. In particular, studies with this fungus may help to resolve the argument as to whether control of differentiation is primarily at the level of transcription or translation.

Studies on the biochemistry of conidiogenesis in filamentous fungi are only now beginning to produce meaningful results. The results of many older studies must be considered with much reservation because of the complex nature of the experimental systems. Conidiogenesis is a very complex type of development and for this reason great care must be taken to appreciate the role of the environment in controlling this process.

The discovery of microcycle conidiation in *Aspergillus niger* and in other fungi could well be of considerable importance since (*a*) it allows conidiation to occur with a minimum of vegetative growth and consequently less masking of important biochemical events; (*b*) a high degree of synchrony can be induced in large populations of cells; and (*c*) the entire cycle can be carried out in a fermenter where many environmental parameters can be suitably monitored.

ULTRASTRUCTURAL ASPECTS OF CONIDIATION

Although there have been many studies on the metabolic aspects of conidiation, there have been few attempts to thoroughly record the ultrastructural changes associated with this developmental process. In particular, it is of major importance to attempt to integrate studies at a biochemical level with structural and ultrastructural observations. Only by considering both aspects together can a composite understanding of conidiation be forthcoming.

Weiss and Turian (1966) have observed several ultrastructural differences between vegetative and conidiating cultures of *Neurospora*. Mitochondria from conidiating mycelium are much more swollen than mitochondria from vegetative cultures, and ribosomes are freely dispersed in the conidia but grouped in zones in vegetative hyphae. Some differences in wall structure have also been noted.

In the growing conidiophore stalk of *Aspergillus niger* (Oliver, 1972) the organelles of the cell are arranged in a similar manner to those in mycelial hyphae. The apex is singularly free of organelles other than vesicles and further behind is the zone of ribosomes and mitochondria and then nuclei. Vacuolation develops in the older parts of the stalk in association with bodies morphologically similar to autophagic vacuoles.

As the conidiophore tip begins to swell to form the vesicle, the apical vesicles (or microvesicles to distinguish them from the conidiophore vesicle) can be seen randomly dispersed throughout the entire volume of the swollen head. As the vesicle dome matures the microvesicles become concentrated in clusters at the sites of development of the metulae. As the metulae develop apical microvesicles and plasmalemmasomes can be seen associated with the wall surface. Further development of metulae, phialides and conidia is effected by a budding cycle in which a uninucleate growing cell becomes cut off from the adjacent cell and then matures autonomously. Throughout this process microvesicles are always present at the growing apex and are normally aligned against the wall surface rather than clustered near by. Plas-malemmasomes occur more frequently near or attached to the newly forming cross septum. In living conidiophores and hyphae there is an elaborate pattern of streaming and pulsation of organelles. Is it possible that cytoplasmic streaming could be an important factor controlling morphogenesis by selectively placing organelles, in particular vesicles, in their correct location?

Summary

Within the fungi there are many different types of asexual reproduction and certain species can individually develop more than one type depend-ing on the environmental conditions. In general, asexual differentiation commonly occurs after the phase of rapid vegetative growth. The in-compatibility of growth and asexual development has led to the generalization that reproduction is initiated by factors which check growth. Conditions permitting asexual reproduction are almost always narrower in range than those permitting mycelial growth. If growth and asexual differentiation do not occur simultaneously and are separated by a definite metabolic shift, the point of change is likely to be associated with limitation of vegetative growth due to nutrient exhaustion. Several examples of this can be seen in the filamentous fungi.

Advances in understanding the biochemical bases of asexual repro-duction have come by way of experimental systems which afford the maximum control over the developing system. Although submerged liquid cultivation presents the most suitable experimental system it must always be remembered that this is not the normal environment for most fungi. Clarification of the biochemical control aspects may only reflect these particular growing conditions and may bear no resemblance to the normal surface growth pattern encountered in nature. We must always guard against creating artefacts.

Many different environmental factors can influence development. How do these factors achieve their effects? Do they impinge on a com-mon mechanism or are there many different sensory systems which can individually alter metabolism in favour of the new developmental path-way? This is clearly an area that warrants more investigation.

Sclerotium formation is undoubtedly one of the most complex of asexual developments. The regular morphology of the sclerotium sug-gests intercellular interactions. Do these occur by way of physical con-tact stimuli or by metabolite exchange between neighbouring cells? Can hormones be implicated?

The totipotency of fungal cells can be well seen in *Blastocladiella*. Only one avenue of development has been extensively studied, and it would be of considerable value to know what environmental forces are involved in the other cases. What is the exact role of CO_2 in morphogenesis? Current work would suggest that this gas exerts very considerable effects on morphogenesis. It may never be possible to know the exact intercellular concentrations of CO_2 in multicellular tissues.

Information derived from experiments performed with isolated enzyme preparations has been of questionable value. Increasing evidence is accumulating which would indicate that results obtained in conventional enzymatic assays concerning kinetic and regulatory properties may not be directly extrapolable to the actual situation of the cell. Methods must be developed that will allow *in vitro* determination of enzyme activity.

The involvement of cellular vesicles with developmental changes has been demonstrated with many forms of asexual morphogenesis. Without doubt they are of major importance in allowing changing metabolism to be reflected in changing morphology. It has been suggested that cytoplasmic streaming and filament formation may guide the vesicles to the site of new growth. Only by integrating ultrastructural and biochemical studies can we ever hope to achieve a full and comprehensive understanding of these many and varied types of morphogenesis.

6.7 Recommended Literature

General Reviews

Alexopoulos, C. J. (1962). "Introductory Mycology." John Wiley and Sons, New York and London.

Baldwin, H. H. and Rusch, H. P. (1965). The chemistry of differentiation in lower organisms. *Annual Review of Biochemistry* **34**, 565–594.

Butler, G. M. (1966). Vegetative structures. *In* "The Fungi", **2**, 83–112. Ed. Ainsworth, G. C. and Sussman, A. S. Academic Press, London and New York.

Hawker, L. E. (1957). "The Physiology of Reproduction in Fungi". Cambridge University Press, London and New York.

Hawker, L. E. (1966). Environmental influences on reproduction. *In* "The Fungi", **2**, 435–469. Ed. Ainsworth, G. C. and Sussman, A. S. Academic Press, New York and London.

Morton, A. G. (1967). Morphogenesis in fungi. *Science Progress, Oxford* **55**, 597–611.

Smith, J. E. and Galbraith, J. C. (1971). Biochemical and physiological aspects of differentiation in the fungi. *Advances in Microbial Physiology* **5**, 45–134.

Turian, G. (1969). "Differentiation Fongique". Masson et Cie. Paris.

Chlamydospore Formation

Cochrane, V. M. and Cochrane, J. C. (1970). Chlamydospore development in the absence of protein synthesis in *Fusarium solani*. *Developmental Biology* **23**, 345–354.

Cochrane, V. W. and Cochrane, J. C. (1971). Chlamydospore induction in pure culture in *Fusarium solani*. *Mycologia* **63**, 462–477.

Sclerotium Formation

Chet, I. and Henis, Y. (1968). The control mechanism of sclerotial formation in *Sclerotium rolfsii* Sacc. *Journal of General Microbiology* **54**, 231–236.

Chet, I., Retig, N. and Henis, Y. (1972). Changes in total soluble proteins and in some enzymes during morphogenesis of *Sclerotium rolfsii*. *Journal of General Microbiology* **72**, 451–456.

Okon, Y., Chet, I. and Henis, Y. (1972). Lactose-induced synchronous sclerotium formation in *Sclerotium rolfsii* and its inhibition by ethanol. *Journal of General Microbiology* **71**, 465–470.

Trevethick, J. and Cook, R. C. (1971). Effects of some metabolic inhibitors and sulphur-containing amino acids on sclerotium formation in *Sclerotium rolfsii*, *S. delphinii* and *Sclerotinia sclerotiorum*. *Transactions British Mycological Society* **57**, 340–342.

Willetts, H. J. (1972). The morphogenesis and possible evolutionary origins of fungal sclerotia. *Biological Reviews* **47**, 515–536.

Fission and Budding Yeasts

Bartnicki-Garcia, S. and McMurrough, J. (1971). Biochemistry of morphogenesis. *In* "The Yeasts", **2**, 441–491. Ed. Rose, A. H. and Harrison, J. S. Academic Press, London and New York.

Belin, J. M. (1972). A study of the budding of *Saccharanyses eivarium* Beijerinck with the scanning electron microscope. *Antonie van Laewenhoek* **38**, 341–349.

Beran, K. (1968). Budding of yeast cells: their scars and ageing. *Advances in Microbial Physiology* **2**, 143–169.

Cabib, E. and Bowers, B. (1971). Chitin and yeast budding: localisation of chitin in yeast bud scars. *Journal of Biological Chemistry* **246**, 152–159.

Cabib, E. and Farkas, V. (1971). Control of morphogenesis: an enzymatic mechanism for the initiation of septum formation in yeast. *Proceedings of the U.S. National Academy of Sciences* **68**, 2052–2060.

Halvorson, H. O., Carter, B. L. A., Tauro, P. (1971). Synthesis of enzymes during the cell cycle. *Advances in Microbial Physiology* **6**, 47–106.

Mitchison, J. M. (1971). "The Biology of The Cell Cycle". Cambridge University Press, London and New York.

Mortimer, R. J., and Johnston, J. R. (1959). Life span of individual yeast cells. *Nature, London* **183**, 1751–1752.

Biogenesis of motile spores

Cantino, E. C. (1961). The relationship between biochemical and morphological differentiation in non-filamentous aquatic fungi. *Symposium Society General Microbiology* **11**, 234–271.

Cantino, E. C. (1966). Morphogenesis in aquatic fungi. *In* "The Fungi", **2**, 283–337. Ed. Ainsworth, G. C. and Sussman, A. S. Academic Press, New York and London.

Cantino, E. C. (1967). Dynamics of the point of no return during differentiation in *Blastocladiella emersonii*. *In* "The Molecular Aspects of Biological Development". *National Aeronautical and Space Administration Contract Report* 673, pp. 149–164.

Gay, J. L. and Greenwood, A. D. (1966). Structural aspects of zoospore production in *Saprolegnia ferax* with particular reference to the cell and vascular membranes. *In* "The Fungus Spore", pp. 95–110. Ed. Madelin, M. F. Butterworth, London.

Goldstein, A. and Cantino, E. C. (1962). Light-stimulated polysaccharide and protein synthesis by synchronized, single generations of *Blastocladiella emersonii*, *Journal of General Microbiology* **28**, 689–699.

Khuow, B. T. and McCurdy, H. D. (1969). Tricarboxylic acid cycle enzymes and morphogenesis in *Blastocladiella emersonii*. *Journal of Bacteriology* **99**, 197–205.

Lovett, J. S. and Cantino, E. C. (1960). The relation between biochemical and morphological differentiation in *Blastocladiella emersonii*. *American Journal of Botany* **47**, 499–505.

McCurdy, H. D., Jr., and Cantino, E. C. (1960). Isocitratase, glycine-alanine transaminase, and development in *Blastocladiella emersonii*. *Plant Physiology* **35**, 463–476.

Biogenesis of non-motile Spores

Anderson, J. G. and Smith, J. E. (1971*a*). Synchronous initiation and maturation of *Aspergillus niger* conidiophores in culture. *Transactions of the British Mycological Society* **56**, 9–29.

Anderson, J. G. and Smith, J. E. (1971*b*). The production of conidiophores and conidia by newly germinated conidia of *Aspergillus niger* (microcycle conidiation). *Journal of General Microbiology* **69**, 187–197.

Anderson, J. G. and Smith, J. E. (1972). The effects of elevated temperature on spore swelling and germination in *Aspergillus niger*. *Canadian Journal of Botany* **18**, 289–297.

Bent, K. J. and Morton, A. S. (1964). Amino acid composition of fungi during development in submerged culture. *Biochemical Journal* **92**, 260–269.

Bianchi, D. E. and Turian, G. (1967). The effect of nitrogen source and cysteine on the morphology, conidiation and a cell-wall fraction of conidial and aconidial *Neurospora crassa*. *Zeitschrift für Mikrobiologie* **7**, 257–263.

Bracker, C. E. (1968). The ultrastructure and development of sporangia in *Gilbertella persicaria*. *Mycologia* **60**, 1016–1067.

Carter, B. L. A. and Bull, A. T. (1969). Studies of fungal growth and intermediary carbon metabolism under steady and non-steady state conditions. *Biotechnology and Bioengineering* **11**, 785–804.

Dicker, J. W., Oulevey-Matikian, N. and Turian, G. (1969). Amino acid induction of conidiation and morphological alterations in wild type and morphological mutants of *Neurospora crassa*. *Archiv für Mikrobiologie* **65**, 241–257.

Ellis, M. B. (1971). "Dematiaceous Hyphomycetes". Commonwealth Mycological Institute, London.

Galbraith, J. C. and Smith, J. E. (1969). Sporulation of *Aspergillus niger* in submerged liquid culture. *Journal of General Microbiology* **59**, 31–45.

Galun, F. and Gressel, J. (1966). Morphogenesis in *Trichoderma*: Suppression of photoinduction by 5-fluorouracil. *Science* **151**, 696–698.

Gooday, G. W. (1968). The localization of some enzymes in the mycelium of *Mucor hiemalis*. *Archiv für Mikrobiologie* **63**, 11–14.

Hawker, L. E. and Hepden, P. M. (1962). Sporulation in *Rhizopus sexualis* and some other fungi following a period of intense respiration. *Annals of Botany* **26**, 619–632.

Lloyd, G. I., Anderson, J. G., Smith, J. E. and Morris, E. O. (1972). Conidiation and esterase synthesis in *Aspergillus niger*. *Transactions of the British Mycological Society* **59**, 63–70.

Morton, A. G. (1961). The induction of sporulation in mould fungi. *Proceedings of the Royal Society B* **153**, 548–569.

Ng, W. S., Smith, J. E. and Anderson, J. G. (1972). Changes in carbon catabolic pathways during synchronous development in *Aspergillus niger*. *Journal of General Microbiology* **71**, 495–504.

Ng, A., Smith, J. E. and McIntosh, A. F. (1972). Conidiation of *Aspergillus niger* in continuous culture. *Archiv für Mikrobiologie* **88**, 119–126.

Oliver, P. T. P. (1972). Conidiophore and spore development in *Aspergillus nidulans*. *Journal of General Microbiology* **73**, 45–54.

Raper, K. B. and Fennell, D. I. (1965). "The Genus Aspergillus". Williams and Wilkins Co, Baltimore.

Righelato, R. C., Trinci, A. P. J., Pirt, S. J. and Peat, A. (1968). The influence of maintenance energy and growth rate on the metabolic activity, morphology and conidiation of *Penicillium chrysogenum*. *Journal of General Microbiology* **50**, 399–412.

Smith, J. E. and Anderson, J. G. (1973). Differentiation in the *Aspergilli*. *Symposium Society of General Microbiology* **23**, 295–337.

Smith, J. E. and Ng, W. S. (1972). Fluorometric determination of glycolytic intermediates and adenylates during sequential changes in replacement culture of *Aspergillus niger*. *Canadian Journal of Microbiology* **18**, 1657–1664.

Stavy, R., Galun, R. and Gressel, J. (1972). Morphogenesis in *Trichoderma*: RNA. DNA hybridization studies. *Biochimica et Biophysica Acta* **259**, 321–329.

Stine, G. J. and Clark, A. M. (1967). Synchronous production of conidiophores and conidia of *Neurospora crassa*. *Canadian Journal of Microbiology* **13**, 447–453.

Turian, G. (1966). Morphogenesis in Ascomycetes. *In* "The Fungi" **2**, 339–385. Ed. Ainsworth, G. C. and Sussman, A. S. Academic Press, New York and London.

Turian, G. (1966). The genesis of macroconidia of *Neurospora*. *In* "The Fungus Spore", pp. 61–67. Ed. Madelin, M. F., Butterworths, London.

Turian, G. and Bianchi, D. E. (1971). Conidiation in *Neurospora crassa*. *Archiv für Mikrobiologie* **77**, 262–274.

Weiss, B. and Turian, G. (1966). A study of conidiation in *Neurospora crassa*. *Journal of General Microbiology* **44**, 407–418.

Yanagita, T. and Kogane, F. (1962). Growth and cytochemical differentiation of mold colonies. *Journal of General and Applied Microbiology* **8**, 201–213.

7 Sexual Reproduction

7.1 Introduction

Sexual processes in the fungi may be defined as those processes necessary in achieving the juxtaposition and fusion of compatible haploid nuclei together with the recombination of genetic factors by meiosis. These are alternate processes in which plasmogamy brings the two nuclei together in one cell, karyogamy unites them into one diploid, and meiosis subsequently re-establishes the haploid state.

In the higher Ascomycetes and in the Basidiomycetes, the processes of plasmogamy and karyogamy may be separated in time and space, and nuclear fusion may not take place until later in the life history of the fungus. This distinctive, binucleate, haploid phase in a life cycle is called the dikaryon and serves to increase the eventual productivity and possible genetic combinations per sexual fusion. The process of dikaryon formation is initiated by the fusion of two sexually compatible elements, spores, vegetative cells or highly differentiated sexual organs each containing one or more haploid nuclei. After plasmogamy, the two nuclei retain their individuality and become associated in pairs which may propagate for a short or indefinite period of time by simultaneous mitotic division (Fig. 7.1). This process is termed conjugate division, and apart from the Myxomycetes is the only other example within the fungi where there is consistent synchronous nuclear division within a cell. It is impossible at this stage to gain an insight into the factors controlling synchrony, since only adjoining nuclear pairs and not all of the nuclei in a filament are synchronous.

A wide range of life cycles exist within the sexually reproducing fungi, extending from completely haploid at one extreme, to completely diploid, except for immediate products of meiosis, at the other. The seven main types of life cycle found in the fungi are diagrammatically represented in Fig. 7.2 (Raper, 1966).

A. *Asexual cycle*. Fungi in this group lack overt sexuality and propagate themselves by continued asexual means. However, some genera

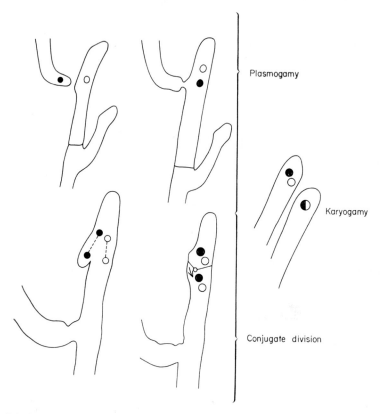

Fig. 7.1. Schematic representation of the initiation, propagation, and termination of the dikaryotic association of compatible nuclei. The dikaryon occurs in the higher Ascomycetes in the ascogenous hyphae and universally in the Basidiomycetes in the "secondary" (dikaryotic mycelium). From Raper (1966).

can achieve somatic recombination by way of the parasexual cycle. The *Fungi Imperfecti* are the main examples of this cycle and include most of the industrially important filamentous fungi such as *Penicillium*, *Aspergillus* and *Cephalosporium*.

B. *Haploid Cycle.* This cycle is found in most Phycomycetes and some Ascomycetes and is completely haploid with the exception of a single diploid nuclear generation, the zygote. Because of its relative simplicity and ease of control, this cycle has been extensively studied at a physiological and biochemical level.

C. *Haploid Cycle with Restricted Dikaryon.* This condition occurs in some Ascomycetes, notably *Neurospora*, and involves a short period after plasmogamy when a dikaryotic condition prevails. The dikaryotic phase

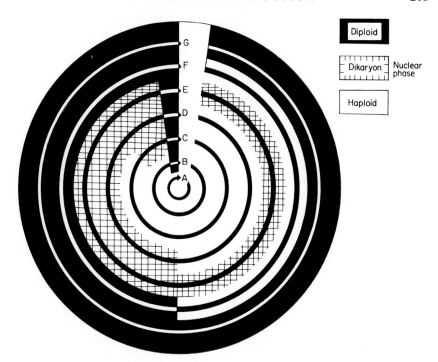

Fig. 7.2. Schematic comparison of life cycles in fungi. In each cycle, changes in nuclear phase are indicated, in clockwise sequence, by changes in shading. The double vertical line at the top of the diagram represents meiosis, and each of the two narrow sectors adjacent to the line represents a single nuclear generation. From Raper (1954).

is totally dependent on the haploid mycelium and cannot exist alone.

D. *Haploid—Dikaryotic Cycle*. This is the condition that prevails in most of the Basidiomycetes in which there is an unrestricted and independent growth phase of the dikaryon. The monokaryotic and the dikaryotic phases are each capable of independent existence and normally require differing physiological conditions. The clear separation of the two phases has permitted extensive experimental examination of the factors regulating the initiation and development of each state, particularly in *Schizophyllum* and *Coprinus*.

E. *Dikaryotic Cycle*. In this cycle, found in some yeasts and smuts, the haploid and true diploid phases are reduced to single nuclear generations and the prolonged growth phrase is the dikaryon.

F. *Haploid-Diploid Cycle*. In fungi such as *Allomyces* true alternation of haploid and diploid generations exists. The occurrence of the phases on separate plants has again allowed extensive biochemical examination.

G. *Diploid Cycle.* This is the condition which prevails in the higher plants and animals where the cycle is completely diploid except for the immediate products of meiosis. Examples of this cycle in the fungi are found in some yeasts and the Myxomycetes.

Undoubtedly, an important feature of sexual reproduction is the occurrence of meiotic nuclear division at some point in the life cycle of the fungus. In most fungi, the occurrence of meiosis results from preceding morphological events such as fusion of nuclei or germination of the zygote while in certain diploid strains of yeast meiosis can be directly induced by changes in environmental conditions.

Chemical communication between individual fungi has been postulated for many years. The induction of chemotaxis and chemotropism by hormone-like secretions has been demonstrated but rarely fully characterized. However, it is now widely accepted that there are definite physiological mechanisms independent of the genes controlling incompatibility, which govern sexuality. There is growing evidence that the secretion of hormones controls and directs the initiation of sexually active organs or cells in many fungi. Sexual hormones have been defined as transported chemical substances which regulate sexual reproduction up to but not including the actual union of gametes or gametangia. The structure of only three fungal sex hormones has been

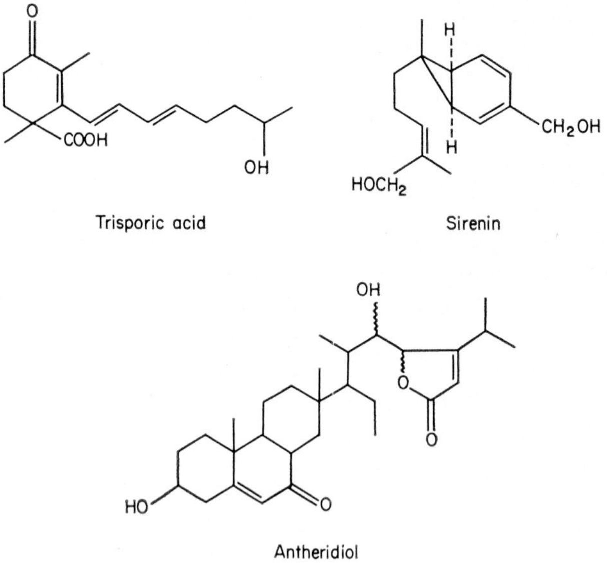

Trisporic acid

Sirenin

Antheridiol

Fig. 7.3. Chemical structure of three compounds functioning in sexual reproduction in Phycomycete fungi. All three are terpenoid in character.

characterized up till now (Fig. 7.3); sirenin, a sperm attractant in *Allomyces*, and two which induce sexual structures, antheridiol from *Achlya* and trisporic acid from *Mucor* and *Blakeslea*. With few exceptions, little is known about the physiological and biochemical changes that occur within an organism during sexual morphogenesis.

Although only three distinct fungal hormones have as yet been chemically categorized, Machlis (1972) has suggested that generic terms should be given to denote the type of sexual activity promoted by the hormone. Hormones inducing tactic movements such as sirenin are erotactins; those inducing and controlling the differentiation of sexual structures, such as antheridiol and trisporic acid, are erogens, while hormones controlling chemotactic growth are erotropins. Such terminology will have obvious advantages as more hormone-like compounds are isolated and chemically characterized.

7.2 Sexual Reproduction in the Phycomycetes

The three chemically characterized fungal sex hormones are all derived from Phycomycete fungi. Almost all investigations have been concerned with establishing their biological role, methods of isolation and chemical characterization. Because of the difficulty in obtaining sufficiently large quantities of these hormones, there have as yet been few studies on the metabolism of the hormones and on their mechanism of action.

Class Chytridiomycetes

The main distinguishing feature of fungi in this group is the production of motile cells each with a single, posterior, whiplash flagellum. In species of *Allomyces*, e.g. *A. arbuscula* and *A. macrogynus*, sexual reproduction is accomplished by the copulation of free swimming anisogamous planogametes in a liquid environment (Fig. 7.4). The male gamete is orange in colour due to the presence of y-carotene and is about half the size of the colourless, female gamete. The gametes are produced in separate gametangia on the same thallus. Prior to their release the female gametes secrete into the environment a sperm attractant—sirenin. Male gametes cluster around the female gametangia and fusion results as the female gametes emerge. The product of fusion is a motile biflagellate zygote which ultimately settles down and produces a diploid vegetative plant or thallus.

To facilitate the isolation and fractionation of sirenin from female gametangia, the normal hermaphroditic growth form of the fungus was

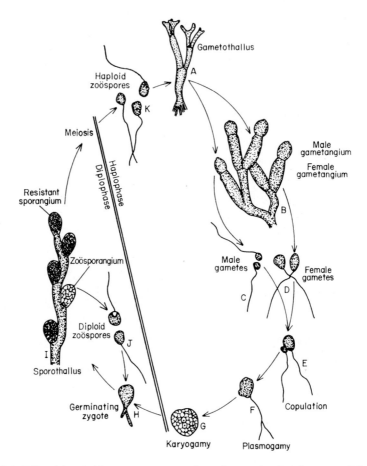

Fig. 7.4. Life cycle of *Allomyces macrogynus*. Sexual reproduction is accomplished by the copulation of free swimming anisogamous planogametes in a liquid environment. From Alexopoulos (1962).

bypassed by the selection of interspecific hybrids in which 96–99% of the gametangia were either female or male. Sirenin was extracted with methylene chloride from an aqueous phase in which large numbers of female gametes had been present. The method of collection and purification was extremely difficult, but it is now possible to synthesize sirenin from relatively inexpensive precursors and this should allow a wider use of this hormone for biochemical studies. The bioassay of sirenin makes use of a miniature glass vessel with a membraneous base. The sirenin solution is placed in the vessel and the number of male gametes congregating beneath the membrane can be measured. The response of the male gamete extends to 10^{-10} M and increases up to 10^{-6} M. Above this, stimulation levels-off due probably to the saturation of the chemoreceptors. The production, isolation, characterization and chemical structure of sirenin has been described by Machlis, Nutting and Rapaport (1968). Sirenin is an oxygenated sesquiterpene and there may be several species specific sirenins. Both D- and L-sirenins have been isolated, although the D-form has no hormone activity.

Future studies with this hormone should illuminate the means by which the male gamete detects the hormone and also what happens to the hormone once it has entered the gamete.

Class Oömycetes

Sexual reproduction is almost invariably heterogametangic and the formation of oöspores is characteristic of all but the most primitive species. The sex organs which are morphologically distinct consist of a tubular or lobed male structure, the antheridium, and a spherical female organ, the oögonium. Sexual reproduction, involving the production and fusion of the sex organs, is initiated and regulated by hormones that appear to be sequentially secreted by the sexual partners into their aqueous surroundings.

ORDER SAPROLEGNIALES

Sexual reproduction involves the development either on the same thallus (homothallic) or on separate thalli (heterothallic) of spherical oögonia with multinucleate eggs or oöspheres and cylindrical antheridial hyphae. The sexual reproductive organs are rarely intercalary, forming normally at the tips of hyphae and hyphal branches. Fertilization is by means of gametangial contact, male gametes passing into the female gametangium through a fertilization tube. Most investigations have been carried out on *Achlya* and the major findings have been comprehensively reviewed by Barksdale (1969).

The first definite proof that sexual hormones were functioning in *Achlya* was given at least two decades ago by Raper. In a classic series of experiments, Raper showed that sexual reproduction in carefully chosen species of *Achlya* was initiated and controlled by diffusible substances reciprocally secreted by the sexually reacting pair of thalli (Fig. 7.5).

When a filtrate from female thalli was added to a culture of male thalli antheridial initials were induced on the male. If the water in which the male hyphae with antheridia was later filtered and the

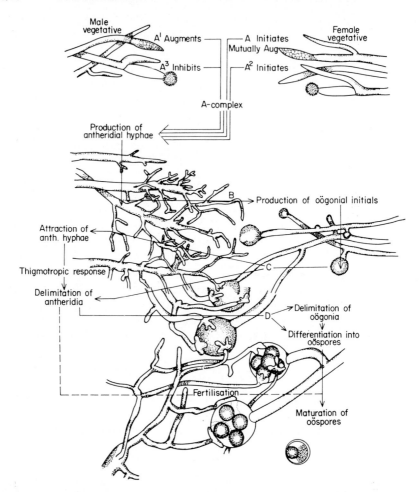

Fig. 7.5. A semidiagrammatic representation of the sexual progression in hetero-thallic species of *Achlya* relating the sequences of morphological developments to the origins and specific activities of the several sexual hormones. From Raper (1955).

filtrate added to the female plants, oögonial initials developed on the female. It was considered that a secretion from the vegetative hyphae of the female plants initiated the sexual reaction by inducing the formation of the antheridial branches and subsequently the now sexually aroused male secreted a substance that induced the formation of oögonial initials on the female. The female secretion was called hormone A and the male secretion hormone B. A third hormone, hormone C, was considered to attract the antheridial branches and delimit the antheridia. Finally, it was considered that the cleavage which gave rise to the oösphere in the female oögonium was controlled by a fourth hormone, hormone D.

Hormone A has now been isolated and characterized and is called antheridiol. Antheridiol is active on receptive strains at concentrations down to 10^{-10} M and the pure hormone elicits a sequence of distinct responses from the male plants of branching, chemotropic attraction, and delimitation of the antheridia with associated meiosis. This is the first steroidal sex hormone to be recognized in fungi, and also, for that matter, in plants, and differs structurally from the mammalian sex hormones in that it has a much longer side chain attached at C-17. The chemical synthesis of antheridiol had not yet been achieved. It has also been suggested that antheridiol may also be hormone C. Hormone B has been extracted from the culture liquid but has not yet been isolated in pure form.

ORDER PERONOSPORALES

Sexual reproduction is again by way of well developed oögonia and antheridia borne on the same or different hyphae, and represents the highest development achieved in the class Oömycetes. Sexual hormones have not been isolated from this order. However, in some species sterols appear to be essential for sexual reproduction, in particular *Pythium* and *Phytophthora*. These fungi do not appear to be able to enzymatically synthesize sterols, and although vegetative growth does not have an obligate sterol requirement, sexual reproduction is greatly enhanced in the presence of sterols. The connection between the sterol requirement and the formation of sexual organs is far from clear but it is obvious that sterols are essential for expression of sexuality in at least this group of Oömycete fungi (Table 7·1). Hendrix (1970) has comprehensively reviewed the role of sterols in the growth and reproduction of fungi. However, the question as to how sterols affect morphogenesis in fungi is still unanswered. It is possible that too much emphasis has been placed on the use of inhibitory chemicals and not enough attention has been paid to the biochemical changes occurring during differentiation. It is quite probable that the sterols may be involved in a hormonal control of sexuality.

Class Zygomycetes

In this group of fungi sexual reproduction occurs between morphologically indistinguishable mating types designated $(+)$ and $(-)$ by

Table 7.1. Mean counts of oöspores of *Phytophthora cactorum* produced in media with various sterols. From Elliott (1972).

Each value is the mean of 4 counts in each of 5 Petri dishes, or occasionally of 5 counts in each of 4 dishes. a and b: replications made by supplementing basal medium from two independent serial dilutions of a common stock solution of the sterol.

Experiment	Sterol		Sterol concentrations (mg/l)					
			25	10	4	1·6	0·64	0·256
1	Cholesterol	a	25·6	25·7	30·9	22·4	4·5	2·5
		b	23·5	32·5	15·5	26·1	5·3	2·5
	β-Sitosterol	a	56·3	43·3	66·9	43·8	24·7	3·5
		b	60·8	70·6	78·8	37·3	20·7	9·2
	Fucosterol	a	76·0	67·4	59·6	26·4	8·2	2·2
		b	85·3	86·6	70·8	33·4	10·1	1·8
2	Cholesterol		16·8	26·6	34·8	23·5	10·8	2·0
	β-Sitosterol	a	160·7	136·3	80·3	31·0	11·3	0·8
		b	134·5	104·5	83·0	28·3	13·3	2·4
	Fucosterol	a	132·3	97·3	82·6	35·2	19·6	2·9
		b	138·0	112·0	79·8	35·7	12·7	2·5
	Ergosterol	a	18·5	28·7	21·8	10·8	4·0	0·2
		b	12·7	30·9	29·3	10·0	2·1	0·2
3	Cholesterol	a	16·9	11·4	9·5	10·2	2·9	0·4
		b	9·4	12·3	11·1	7·3	3·4	1·0
	Stigmasterol	a	52·8	37·5	14·6	11·2	3·4	0·8
		b	30·6	42·7	26·4	12·8	5·5	0·1
	5-Avenasterol		42·9	25·0	31·8	21·5	1·5	1·4
4	Cholesterol	a	101·3	77·8	74·4	48·5	15·3	6·2
		b	77·2	85·9	67·5	34·1	13·0	2·4
	7-Dehydro-	a	102·0	75·1	55·3	30·0	10·6	1·8
	cholesterol	b	116·2	86·9	73·6	22·8	7·5	1·4
5	Cholesterol	a	85·3	71·4	44·4	21·4	12·4	2·3
		b	62·4	54·2	42·8	31·2	14·3	3·1
	Fucosterol	a	203·8	220·3	129·9	57·0	27·7	8·5
		b	190·6	142·5	86·8	61·7	36·1	9·4
	Stigmasterol	a	189·3	157·2	77·2	60·5	17·4	4·1
		b	169·1	139·7	108·6	60·0	21·8	7·6

the copulation of multinucleate haploid gametangia which are mainly similar in structure but may differ in size. Such fungi are said to show physiological heterothallism.

ORDER MUCORALES

Sexual differentiation is initiated by the proximity of compatible mating strains without the need for physical contact. The sexual hyphae or zygophores grow towards each other by a process of zygotropism which appears to be controlled by a volatile hormone. Fusion occurs in mated pairs, and there is rapid swelling at the tip area in each partner to give progametangia (Fig. 7.6). The tip area is delimited by cross-walls to form the gametangia and these normally become clearly supported by distinctive suspensor cells. After the gametangia fuse, the fusion wall breaks down to allow plasmogamy to occur. This new fusion structure, the zygospore, develops a thick, dark resistant wall and at a later date karyogamy and meiosis will occur within the zygospore.

The classical experiments of Burgeff (1924) demonstrated that a diffusible substance(s) was responsible for the formation of the sexual hyphae in certain Mucorales when two opposite mating types were in close proximity to each other. Studies carried out later in the 1950s on *Mucor mucedo* by Plempel and his associates suggested the involvement of many hormones to achieve the full sexual cycle. They showed that the vegetative mycelium of one mating type $(-)$ secretes an erogenic hormone that sexually activates the $(+)$ strain which in turn secretes an erogenic hormone that will induce the formation of zygophores in the $(-)$ strain. An opposite sequence results in the production of zygophores in the $(+)$ strain. Finally, each strain produces an erotropic hormone that directs the growth of the zygophores of the opposite strain. They were able to extract a good quantity of erogen but were unable to characterize it.

An interesting observation with sexual reproduction in all species examined was the development of a yellow colour in the zygophores due to the presence of β-carotene. In particular, mixed sexually active cultures of *Blakeslea trispora* produced major quantities of carotene. A soluble factor present in mated cultures of *B. trispora*, later identified as trisporic acid, caused a large increase in carotenogenesis when added to unmated strains. Subsequent examination of this soluble factor showed that it contained trisporic acids A, B and C with C constituting about 80%, B, 15% and A, 1–2%. The erogenic factors from other Mucorales have also been identified and shown to be trisporic acids B and C. Thus it would seem almost certain that the trisporic acids induce zygophores in both mating types of at least several species of the Mucorales.

The trisporic acids are not produced by single strains of the organisms, but only by mixed cultures. The exact mechanism of this synergistic biosynthesis is not yet clear. ^{14}C studies suggest an almost equal co-operation, and inhibitor studies with 5-fluorouracil indicate a need for RNA-mediated enzyme synthesis in both strains for continued trisporic acid formation. It is now clear that β-carotene is a precursor of trisporic

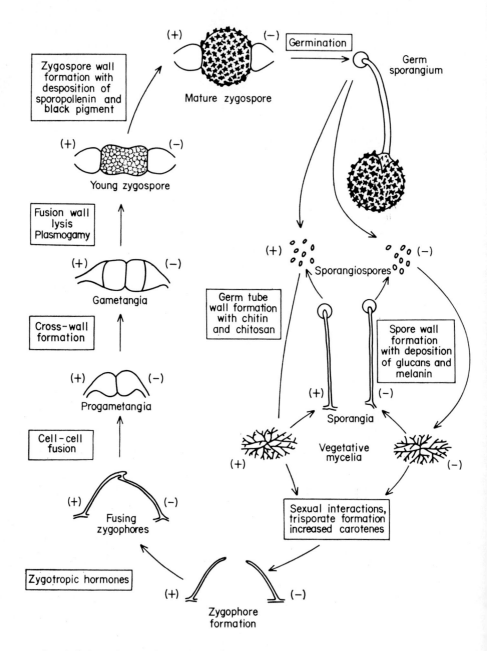

Fig. 7.6. Sexual reproduction in the Mucorales. The sexual hyphae or zygophores grow towards each other and fusion occurs in mated pairs. From Gooday (1973).

acid biosynthesis and that trisporic acid itself then acts to stimulate β-carotene production in the zygophores. Radio-active tracer studies have shown that the carotene formed during sexual reproduction is oxidatively polymerized to give sporopollenin, a component in the highly resistant zygospore wall.

The trisporic acids can be rapidly bioassayed by adding a small amount of the hormone into a well just in front of the advancing hyphal tips of $(+)$ or $(-)$ cultures growing on nutrient agar. A few hours later, zygophores can be observed around the zone of diffusion and the relationship between number of zygophores and concentration of hormones can be assessed. The zygophores and their associated vegetative mycelium are a much brighter orange than the mycelium in untreated cultures due to the presence of carotene.

The formation and action of trisporic acid can be summarized as follows (Gooday, 1973):

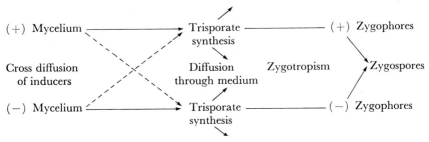

Thus the sexual process in the Mucorales is governed by the initiation of trisporate synthesis, followed by zygophore formation, zygotropism and zygophore fusion. This sequence occurs in an orderly manner and is co-ordinated by inter-cellular chemical communication by the interchange of trisporate inducers, by the diffusion of trisporic acid itself, by the interchange of zygotropic hormones, and by the mutual cell-cell recognition at the two cell surfaces.

In conclusion it can be noted that there is no marked similarity in structure, biochemistry or biological activity between the three fungal sex hormones that have so far been characterized (Table 7.2). The activity of each hormone is limited to closely related species. The success in producing pure quantities of sirenin, antheridiol and trisporic acid has largely been the result of using carefully selected strains of fungi, controlled fermentations and highly sensitive bioassays.

TABLE 7.2. Fungal sex hormones (Gooday, 1973).

	Mucor	Achlya	Allomyces
Hormone	Trisporic acid C	Antheridiol	Sirenin
Molecular formula,	$C_{18}H_{26}O_4$	$C_{29}H_{42}O_5$	$C_{15}H_{24}O_2$
Molecular weight	306	470	236
Optimal yield of hormone (molar concentrations)	5×10^{-4} (*B. trispora*) 5×10 (*M. mucedo*)	6×10^{-9} (*A. bisexualis*)	1×10^{-6} (*A. sarbuscula/ A. jovanicus* hybrid)
Sensitivity of bioassay (molar concentrations)	1×10^{-8}	1×10^{-11}	1×10^{-10}
Specificity of production and of activity	Many families in the Mucorales	*Achlya* spp.	*Allomyces* spp.
Control of synthesis	Interaction between $(+)$ and $(-)$ mating types gives synthesis by $(+)$ and $(-)$ mycelia	Synthesized by female mycelium	Synthesized by female motile gametes
Morphogenetic action of hormone	$(+)$ and $(-)$ mycelia produce zygophores. Sporangiophore production prevented	Male mycelium branches; antheridia de-limited with meiosis; chemotropism of antheridia	Chemotaxis of male gametes
Biochemical action of hormone	Increase in carotenoids and sterols	Increase in cellulase; production of hormone B	

7.3 Sexual Reproduction in the Ascomycetes

Introduction

The Ascomycetes are characterized by the production of asci during sexual reproduction. These asci contain either four ascospores which are the direct products of meiosis, or eight ascospores which are the product of a post-meiotic mitosis. The Ascomycetes can be conveniently divided into the Hemiascomycetes in which the asci are produced singly without a sporophore, and the Euascomycetes in which several asci are produced from ascogenous hyphae in a sporophore. The major group of Hemiascomycetes are the yeasts which constitute a particularly valuable group of organisms for studies on sexual reproduction. Yeasts have a relatively simple life cycle (Fig. 7.7). Haploid cells of appropriate mating types conjugate to form a diploid cell which may undergo meiosis immediately, or may reproduce vegetatively, producing a population of diploid cells, each of which can then undergo meiosis when transferred to a suitable medium. The advantage of yeasts for studies on sexual reproduction is that the majority of a population of cells can be induced to enter conjugation or

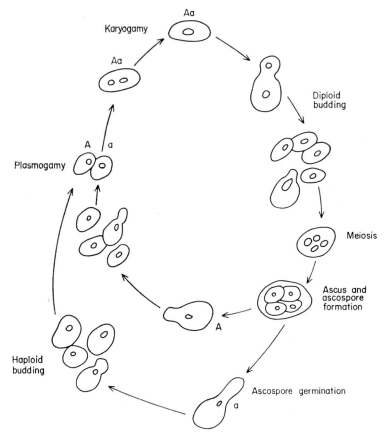

Fig. 7.7. Life cycle of *Saccharomyces cerevisiae*.

meiosis and sporulation at the same time. This is in marked contrast to most organisms where only a few specialized cells of a mycelium are involved in sexual reproduction.

In the Euascomycetes, conjugation is by hyphal fusion (*Aspergillus nidulans*) or by spermatization (*Neurospora crassa*), and the events leading to ascospore formation are usually enclosed within the sporophore. In several species the ascospores are arranged linearly within the ascus. In such an ascus the sequence of ascospores is determined by the orientation of the metaphase plate during meiosis, e.g. *Sordaria fimicola*. Because of this, it is possible to distinguish between first and second division segregation of alleles affecting a character such as spore colour. This system has been used extensively by geneticists for teaching and research purposes (Fincham and Day, 1971).

Hemiascomycetes

PLASMOGAMY IN YEAST

In *Hansenula wingei* a massed agglutination occurs when two strains of opposite mating types are mixed. The agglutination reaction facilitates plasmogamy, and in most strains it is essential if a high rate of plasmogamy and fertilization is to be achieved. However, in some strains a low rate of conjugation occurs in the absence of any obvious agglutination. The mechanism of agglutination and conjugation have been studied in recent years (Crandall and Brock, 1968). Cytological observations indicate that the process can be divided into at least six stages: (1) initial contact, (2) fusion of cell walls, (3) dissolution of cell wall, (4) fusion of cytoplasm, (5) nuclear fusion and (6) formation of a diploid bud (see Fig. 7.8).

After the initial contact of compatible cells, a relatively large area of the cell walls becomes adpressed, indicating a specific binding reaction rather than a random contact of spherical cells. Cell fusion does not follow immediately but is preceded by an extension of the cell wall of both cells in the region of cell contact, providing a structure which could be considered to be a germ tube (b); then the walls which have been in close contact fuse to form a single dividing wall (c). Plasmogamy occurs when this wall is broken down (d). Conjugation appears to be of an isogamous type since the two nuclei come together and fuse at the point of cell fusion (e), and the diploid nucleus produced immediately migrates into a bud which also originates at the point of fusion (f). The movement of nuclei and cytoplasm into the germ tube and later into the diploid bud is accompanied by an increased vacuolation of the original haploid cells (Fig. 7.9).

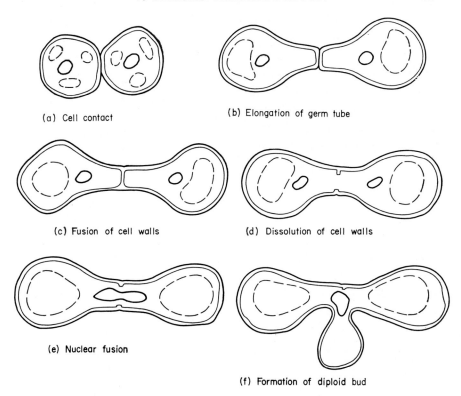

(a) Cell contact

(b) Elongation of germ tube

(c) Fusion of cell walls

(d) Dissolution of cell walls

(e) Nuclear fusion

(f) Formation of diploid bud

Fig. 7.8. Diagrammatic representation of conjugation in *Hansenula wingei*.

The agglutination reaction has been studied using two compatible strains, 5 and 21. An extract has been isolated from the cell wall fraction of strain 5 which causes strain 21 to agglutinate. Ultracentrifuge studies on this extract indicate that the active component varies in size between 18,000 and 200,000 Daltons and that the larger the molecule the more effective it is in inducing agglutination of strain 21. It has been postulated that the larger molecules are aggregates of a monomer and that their increased agglutination activity results from a higher number of binding sites. Chemical analysis has indicated that the active component is a glycoprotein containing 50% carbohydrate, composed mostly of mannose. As yet it has not been ascertained whether the binding site is associated with the carbohydrate or protein fraction of the molecule. Treatment with periodate destroys the agglutinability of the molecule. This action of the periodate is thought to be the primary result of the oxidation of disulphide bridges in the protein moiety rather than the cleavage of mannose molecules. Treatment with

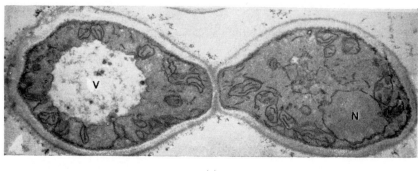

(a)

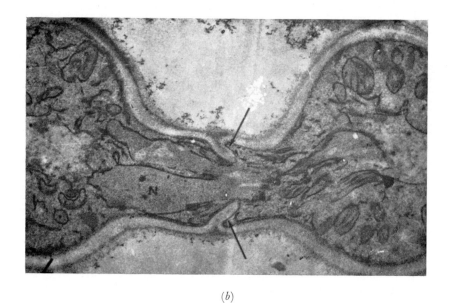

(b)

Fig. 7.9. Electron micrographs of three stages during conjugation in *Hansenula wingei*.

    (a) Cell contact   ×20 000
    (b) Development of the conjugation tube   ×23 700
    (c) Formation of the diploid bud   ×21 000
        V = vacuole   N = nucleus
        Arrows indicate dividing wall
        f.a. = fibrillar material

From Conti and Brock (1965)

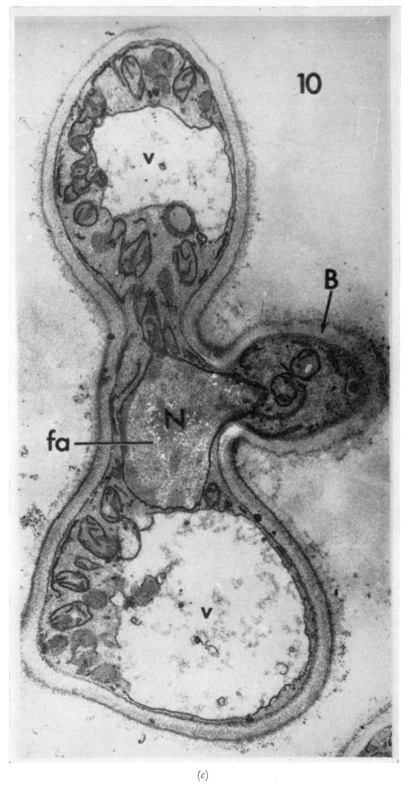

(c)

mannose, which is known to inhibit agglutination in brewer's yeast has no effect on agglutination induced by the factor from strain 5.

Strain 21 does not contain a molecule capable of agglutinating strain 5. However, a molecule has been isolated from the cell-wall of strain 21 which binds the agglutination factor from strain 5 and effectively neutralizes it. Treatment of strain 21 cells with trypsin liberates this factor from the cell-wall and also renders the cells incapable of being agglutinated. This strain 21 factor, which also appears to be a glyco-protein, binds to strain 5 cells. A heat-stable, non-specific factor has also been described. However, its role in agglutination has not been clarified (Crandall and Brock, 1968).

Because the two agglutination factors always segregate with the mating type it has been suggested that the genes responsible are allelic and either closely linked to the mating type locus or identical to it. In view of the marked differences in properties of the two factors, it seems more likely that they are not allelic and that the mating type genes function as regulatory genes controlling the production of the agglutin-ation factors. Evidence has been obtained for the existence of a regulator gene which controls the expression of the gene for agglutination (Herman and Griffin, 1967).

In view of the complexity of the conjugation process, it is not surprising that it is an active process, requiring the expenditure of metabolic energy. The optimal temperature for conjugation is 30°C and an exogenous carbon source is required. Although a nitrogen source is not required, endogenous amino acids and bases are apparently available since both protein synthesis and RNA synthesis appear to be essential. Using different combinations of cycloheximide CXD sensitive and resistant strains, it has been shown that protein synthesis is required in both cells (Table 7.3).

TABLE 7.3. Effect of cycloheximide on plasmogamy in yeasts. From Crandell and Brock (1968).

Strain	a	α	Sporulation on CXD medium
CXD	resistant	sensitive	—
CXD	sensitive	resistant	—
CXD	resistant	resistant	+

Azaguanine inhibits conjugation, but actinomycin D is without effect since it does not appear to penetrate the cell.

In *Hansenula wingei*, cell contact is essential before a mating response is observed, whereas in *Saccharomyces cerevisiae* the mating response appears to be stimulated by a diffusible sex factor. If cells of "a" mating type are grown on nutrient agar which has previously been used to grow cells of "α" mating type, then the cells fail to divide and become elongated. Extracts have been obtained from "α" cells grown in liquid culture which cause elongation of "a" cells, although the reverse is not true. The active component of this extract appears to be an oligopeptide (Duntz et al., 1970). It has also been reported that testosterone and estradiol induce cell expansion in "α" and "a" cells respectively and that these strains excrete a steroid resembling but not identical to testosterone and estradiol (Takao et al., 1970).

MEIOSIS AND SPORULATION IN YEAST

Cytology of Yeast Sporulation. Studies on the cytology of yeast sporulation have been restricted by the fact that under the light microscope the nucleus remains more or less opaque throughout the meiotic division, so the chromosomes are not readily visible. In the cytoplasm, an increase in the degree of vacuolation has been reported which is associated with an increase in the degree of fragmentation of the vacuoles. The number of glycogen and fat granules increases throughout sporulation and these are particularly abundant in the epiplasm produced when the ascospores are delineated within the ascus. A decrease in the basophilic reaction of the cytoplasm during sporogenesis is correlated with a decrease in the level of RNA and ribosomes in the developing asci.

Saccharomyces cerevisiae has also proved a difficult organism to fix for studies in electron microscopy. However, recent studies using osmic acid or glutaraldehyde and osmic acid as the fixatives, combined with the use of serial sections, have led to a closer understanding of the nuclear events occurring during sporogenesis (Moens, 1970; Rapport, 1971; Moens and Rapport, 1971).

The behaviour of the spindle plaques and microtubules during meiosis has been demonstrated by Moens and Rapport (1971) (Fig. 7.10). Unlike those found in mitosis the microtubules only occur on the nuclear side of the nuclear membrane at the onset of meiosis (Fig. 7.11). The spindle plaque is a complex structure consisting of an inner plaque which has two dense bands and a less dense central band, an intermediate zone and an outer plaque which is lightly staining during mitosis and the first division of meiosis but is dense during the second meiotic division. The development of the prospore wall is initiated immediately outside the outer plaque and extends outwards to cover the developing ascospore (Fig. 7.12 a and b). Synaptonemal

Stage I. Spindle plaque is single and not very distinct. The few micro-tubules present on the nuclear side of the membrane are short.

T4-5

a

Stage II. Spindle plaque splits but the two remain in contact through the plaque bridge.

T8

b

Spindle plaques move to face each other. Onset of first division of meiosis.

c

Stage III Spindle plaques move apart, the nucleus elongates but the nuclear membrane remains intact. Extended micro-tubules are visible.

T12

d

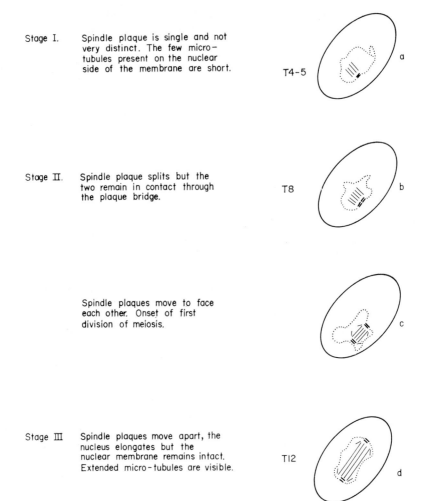

complexes which are considered to be involved in genetic recombination have been described during the first meiotic division in yeast (Fig. 7.13). Genetic evidence indicates that true meiosis occurs during yeast sporulation and, although details of chromosomal behaviour have not been observed, cytological studies appear to confirm this.

Genetics of Yeast Sporulation. Studies on the genetic control of sexual reproduction always present difficulties since the mutants selected frequently impair the sexual function to such an extent that genetic analysis cannot easily be carried out. This can be overcome in yeast by the use of temperature sensitive mutants which sporulate normally

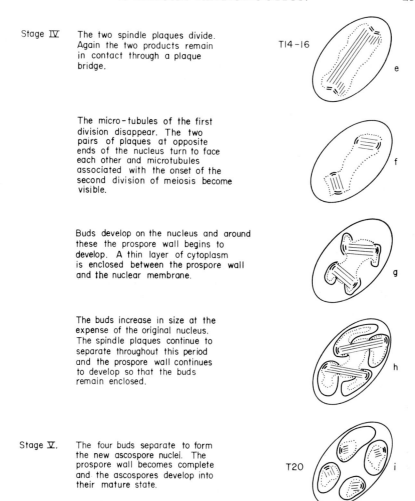

Stage IV   The two spindle plaques divide. Again the two products remain in contact through a plaque bridge.

T14-16   e

The micro-tubules of the first division disappear. The two pairs of plaques at opposite ends of the nucleus turn to face each other and microtubules associated with the onset of the second division of meiosis become visible.

f

Buds develop on the nucleus and around these the prospore wall begins to develop. A thin layer of cytoplasm is enclosed between the prospore wall and the nuclear membrane.

g

The buds increase in size at the expense of the original nucleus. The spindle plaques continue to separate throughout this period and the prospore wall continues to develop so that the buds remain enclosed.

h

Stage V.   The four buds separate to form the new ascospore nuclei. The prospore wall becomes complete and the ascospores develop into their mature state.

T20   i

Fig. 7.10. Diagrammatic representation of meiosis in *Saccharomyces cerevisiae*. After Moens and Rapport (1971)

at the permissive temperature of 25°C but do not sporulate at 34°C (Esposito and Esposito, 1969). The mutants can be isolated and scored at 34°C while genetic analysis is carried out at the permissive temperature.

A further difficulty is that genes affecting meiosis and sporogenesis in yeast are only expressed in the diploid state. Since most of the mutants isolated are recessive, they are not expressed in the normal diploid produced by crossing the mutant strain with a cell of the opposite

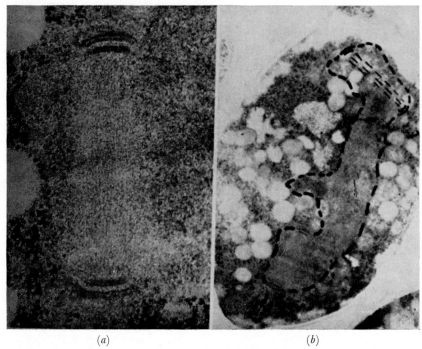

(a)   (b)

Fig. 7.11. Spindle formation during the second division of meiosis in *Saccharomyces cerevisiae*. The limits of the nucleus and one of the spindles are indicated in (b). The spindle in (a) is the same as that shown in (b). The inner and outer plaques are clearly visible. After Moens and Rapport (1971).

mating type because the diploid would be heterozygous for the newly isolated mutant. Techniques have been developed by which homozygous diploids can be produced from haploid mutant strains. In *Schizo saccharomyces pombe*, a mutant, h⁹⁰, has been isolated which is fertile in crosses with h⁺ and h⁻ strains, the normal mating types in *S. pombe* and is self-fertile (Bresch *et al.*, 1968). Mutants induced in this strain can be readily converted to the homozygous diploid state by crossing with sister cells, and the diploid cells produced can then be tested for sporulation. In *S. pombe*, colonies which contain asci can be recognized because they colour black when stained with iodine vapour. Colonies from non-sporulating strains remain unstained (Bresch *et al.*, 1968). A similar technique has been employed with *Saccharomyces cerevisiae*. Two genes HOα and HM$_I$ have been isolated which cause the conversion of the "a" and "α" alleles to the opposite mating type. When this occurs, sister cells are obtained which are fertile and can fuse to form a diploid which is homozygous for all loci except that of the mating type. Using

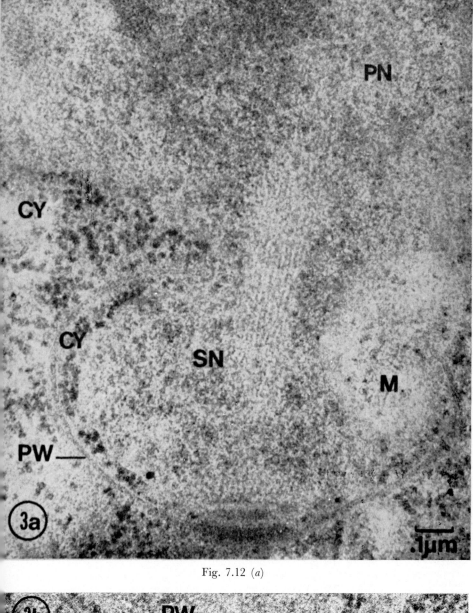

Fig. 7.12 (a)

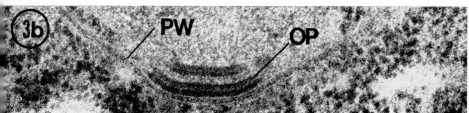

Fig. 7.12 (b)

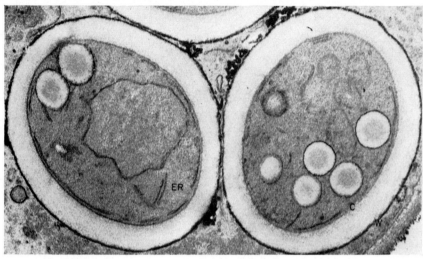

Fig. 7.12 (c)

Fig. 7.12. Ascospore formation.

  (a) Electron micrograph showing prospore wall formation in *Saccharomyces cerevisiae*.
  (b) Electron micrograph showing the intimate relationship between the prospore wall and the outer plaque.
  P.W. = prospore wall
  O.P. = outer plaque
   CY = cytoplasm
  S.N. = spore nucleus
  P.N. = parent nucleus
    M = mitochondrion
  a and *b* after Moens (1971)
  (c) Electron micrograph of an ascus of *Saccharomyces cerevisiae* late in development showing spores of varying degrees of maturity. × 18 000
    W = outer wall
    C = inner coat
   ER = endoplasmic reticulum
  From Lynn and Magee (1970)

this technique, recessive genes which are only expressed in the diploid phase of the life cycle can be identified by converting the haploid strain in which they arise into a diploid which is homozygous for the relevant mutant (Esposito and Esposito, 1969).

   Mutants which affect several different stages in meiosis and sporulation have been isolated using both these techniques. Genetic analysis of the mutants isolated by Bresch (Bresch *et al.*, 1968) led to the identification of 26 complementation groups of which only 3 were sufficiently

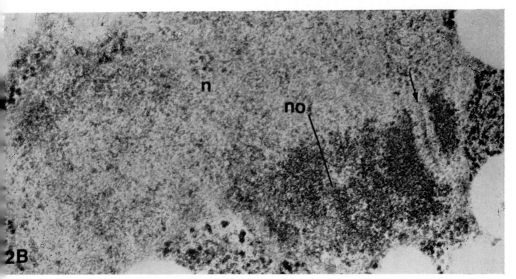

Fig. 7.13. Synaptonemal complexes in *Saccharomyces cerevisiae* (Croes strain No. 10).

n = nucleus
no = nucleus
arrow = synaptonemal complex.

By permission of D. E. Rapport.

closely linked to have any possibility of being within the same gene. It has been suggested that at least 24 genes were involved in sporo-genesis in *S. pombe*. An estimate of 48 ± 27 loci which are involved in sporulation in *S. cerevisiae* has been made by Esposito *et al.*, (1972).

A specific biochemical function has not been allocated to most of the genes recognized, although cytological studies have been carried out to establish at which stage in sporogenesis the genes act. At the present time, mutants have been isolated which affect DNA replication, the first and second division of meiosis and the maturation of the spores (Fig. 7.14).

A specific selection technique has been devised to isolate mutants which have a reduced rate of recombination (Roth and Fogel, 1971). The mating type locus of yeast is found on chromosome III. Using strains which are disomic for chromosome III, that is they are haploid except for an extra copy of chromosome III, it is possible to obtain both "a" and "α" mating types in the same aneuploid cell. Such cells appear to be capable of undergoing meiotic recombination. If two non-complementary auxotrophic markers are incorporated into chromosome III, one on each chromosome, such that a single crossover can give rise to a wild type gene then the frequency of crossovers can

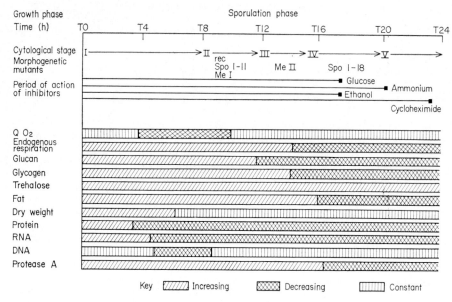

Fig. 7.14. Sequence of biochemical and genetic events during sporulation in *Saccharomyces cerevisiae*. From Tingle *et al* (1973).

be measured by the frequency of prototrophs produced. Strains such as this produce a high frequency of prototrophs when transferred to acetate medium. Since the rest of the genome is haploid, recessive mutations in other chromosomes which affect recombination will be expressed and the frequency of prototrophs decreased or increased. Using ethyl methyl sulphonate as the mutagen, 94 mutants have been isolated which affect the frequency of prototrophs and presumably the frequency of recombination in the disomic chromosomes. These mutants may well affect the DNA polymerase or DNA ligase of the cell.

Other techniques by which mutants for specific stages in sporogenesis could be obtained have recently been put forward (Tingle *et al.*, 1973). These include:

(i) Selection of dominant mutants which are more likely to be associated with regulatory functions.

(ii) Selection of strains which are resistant to those inhibitors which affect sporogenesis.

(iii) Selection for strains which produce abnormal numbers of ascospores per ascus.

(iv) Selection for strains producing ascospores which have an altered sensitivity to heat, osmotic pressure or other physiological constraints.

(v) Studies on mutations which affect vegetative growth in the haploid state and also have a persistent effect on sporulation when crossed with a wild type strain of the opposite mating type.

Physiology of Yeast Sporulation. The most common technique employed for inducing sporulation in *Saccharomyces cerevisiae* is to transfer diploid cells grown in a rich medium containing up to 8 % glucose to a medium lacking nitrogen and containing acetate as the major carbon source. However, not all yeasts sporulate on media such as this which restrict growth; some, e.g. *Debaryomyces* sp., give optimal sporulation on rich media containing meat extract, peptone and glucose (Pfaff *et al.*, 1966). In such species, the termination of mitosis prior to the onset of meiosis must be controlled by specific intracellular mechanisms and cannot be attributed to depletion of exogenous nutrients.

In spite of an immense amount of data available on conditions which favour yeast sporulation, it is still not possible to describe the critical parameters controlling sporulation. In *Saccharomyces cerevisiae*, markedly different conditions appear to be necessary in the presporulation and sporulation phases. However, different investigators have described a wide variety of conditions for both phases which they have found optimal for sporulation (Fowell, 1969).

When yeast is grown aerobically, it achieves an optimal state for transfer to the sporulating medium after 18 h, and any further growth results in a reduction in its capacity to sporulate. However, in anaerobic conditions, cultures grown for up to 48 h are still capable of giving high sporulation rates. Cells grown in a complex medium with glucose as the carbon source have a low capacity for sporulation in the log phase and must be left until the stationary phase before transfer to sporulation medium (Croes, 1967b) whereas those grown on acetate as the carbon source sporulate readily when transferred during the log phase. Conditions which tend to favour sporulation are indicated in Table 7.4. Although acetate is metabolized, it does not appear to be an essential substrate for sporulation, since high levels of sporulation can be achieved by a 5–10 min. exposure to acetate, followed by incubation in water. The removal of the growth medium by centrifugation and its replacement by acetate must constitute a severe physiological shock to the organism. Recently, a technique has been described which allows the growth medium in a fermenter to be gradually replaced by sporulation medium (Watson and Berry, 1973). Using this technique, it should be possible to study the critical events occurring during the transition phase under controlled conditions. Conditions which favour growth tend to inhibit sporulation; for example, the addition of NH_4^+ ions up to 20 h and glucose up to 16–18 h. The reversal of glucose inhibition by

TABLE 7.4. Optimal conditions for sporulation in *S. cerevisiae*

	Presporulation phase	Sporulation phase
Carbon source	Glucose 4–8% or other easily metabolised carbohydrates. Acetate has been used	Acetate 0·1–1·0% dihydroxy-acetone
Nitrogen source	Complex nitrogen sources, yeast extract, meat extract, peptone, tomato juice	Absent or occasionally low levels of amino acids or peptone
Vitamins	β complex vitamins required	Possibly pantothenate
Other constituents	IAA and GAA have been reported to stimulate	CaSO₄ and Na⁺ ions reported to stimulate
O₂	Not essential	Essential, 20% optimal
CO₂	Essential, high concentrations desirable	Low levels required inhibitory above 5%
Temperature	20–30°C	25–30°C
pH	Not critical	5·0–11·0 depending upon strain 7·0–8·0 usual
Other parameters		Cell density 10⁷ cells or less in liquid culture Blue and green light have been reported to inhibit

the addition of cyclic AMP suggests that catabolite repression may be important in controlling yeast sporulation. The addition of ethanol also inhibits sporulation if added before 16 h. This is interesting since alcohol normally accumulates during the log phase when yeast is grown in high concentrations of sugar. It has been postulated that the conversion of alcohol into acetate during the stationary phase is critical in the induction of yeast sporulation (Croes, 1967*b*). Sporulating cells which have been inhibited by the addition of glucose, ammonium ions or ethanol are able to revert to vegetative growth. However, after 16–20 h, the cells are committed to sporulation, and will produce mature asci independently of the addition of these inhibitors.

The percentage of cells producing asci does not normally reach 100% even when logarithmically growing cells are transferred to the sporu-lating medium. Yeast cells show a difference in their capacity to sporulate depending upon the stage of the life cycle which they have reached when they are transferred to the sporulating medium. Daughter cells—those recently derived from buds which do not bear any bud scars—sporulate poorly. Propensity for sporulation increases from the time of bud separation to the time of initiation of the next daughter cell (Haber and Halvorson, 1972).

Biochemical Changes during Sporulation. The changes in the chemical composition of the yeast cell during ascus formation can be related to the different developmental processes which are occurring within the cell. These are meiosis, ascospore formation and the mobilization of endogenous reserves. When compared with vegetative cells, mature asci are richer in glucan, mannan, trehalose and lipids but contain less RNA, protein and polyphosphate. The ascus as a whole naturally contains exactly twice as much nuclear DNA as the vegetative cell.

Sporulation in yeast is a biphasic process. The first phase which must include the induction of sporulation begins in the stationary phase of vegetative growth or earlier and involves the accumulation of metabolites which are then utilized in the second phase of sporulation. The transition from phase one to phase two is not clear-cut, but occurs between 4–12 h after transfer to the sporulation medium (Tingle *et al.*, 1973).

Glucan, glycogen and trehalose, the main storage carbohydrates, begin to accumulate as the growth rate decreases. Whereas glycogen and glucan reach a maximum at about T_{14} (see Fig. 7.14) and then decrease during the sporulation phase, trehalose accumulates continuously up to T_{24} and does not decrease. It appears that glucan and glycogen are the storage carbohydrates which are utilized during sporulation and that trehalose is the storage carbohydrate of the mature ascospores. Mannans, which are structural rather than storage carbohydrates accumulate throughout sporulation.

Fats appear to function as a reserve energy source for ascospores. They accumulate up to T_{20} and are found in the ascospores, rather than the epiplasm of mature asci. The lipid content increases during stages I and II to four times the level found in stationary cells. Most of this increase can be accounted for by increases in the sterol esters and triglycerides (Illingworth *et al.*, 1970). This may be correlated with the observation that the ascospore wall has a superficial lipid layer which is absent in vegetative cells.

DNA replication occurs during stage II immediately before meiosis. Since growth does not occur in the sporulation medium, the percentage increase in DNA gives a measure of the percentage of cells undergoing meiosis (Fig. 7.15). During spore development, it has been reported that the synthesis of mitochondrial DNA and nuclear DNA are synchronized. If this is so, then the number of mitochondrial genomes per ascospore will be half that occurring in the resting stage diploid cell.

The level of RNA and protein in resting phase cells prior to transfer to the sporulation medium is higher than in exponentially dividing cells (Croes, 1967a and b). This increase in the level of RNA and protein only continues until T_4 after which the level of both of them decreases (Fig. 7.15). These results do not, however, indicate that RNA and protein synthesis are not required for sporulation in yeast.

A summary of the information discussed in this section is given in Fig. 7.14.

Metabolism during Sporulation. Several lines of evidence indicate that oxidative phosphorylation is essential for yeast sporulation, and although the total rate of oxygen uptake decreases, the RQ drops from

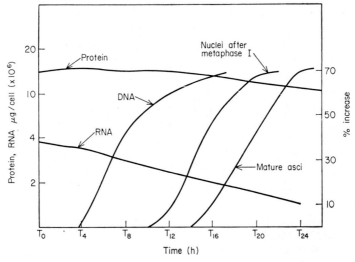

Fig. 7.15. Levels of DNA, RNA and protein during sporulation in yeast. After Croes (1967).

2·6 to 1·5 upon transfer to the sporulation medium. This probably indicates a switch from a fermentative to an oxidative form of energy generation. Uncoupling agents such as 2,4-dinitrophenol inhibit sporulation, as do inhibitors of mitochondrial protein synthesis such as erythromycin and chloramphenicol. The effect of these antibiotics may, however, be on the synthesis of enzymes involved in the TCA cycle and on the glyoxylate cycle rather than the electron transport chain. Oxidative respiration deficient mutants "petites" cannot sporulate.

Both exogenous and endogenous substrates are metabolized during sporulation. Acetate is metabolized at a high rate in stage I, but the rate decreases during the later stages of sporulation whereas the rate of metabolism of endogenous reserves increases up to the beginning of stage V. The period of rapid acetate metabolism is accompanied by an increase in the level of the enzymes of the tricarboxylic acid cycle and of isocitrate lyase, the controlling enzyme of the glyoxylate cycle (Miyake *et al.*, 1971). Of the acetate metabolized by the cell, 62% is lost in respiration, 16% passes into lipids, nucleic acids and proteins and the remainder is found in the pool of intermediates. Experiments using methyl (C_2) and carbonyl (C_1) labelled acetate show that the carbonyl is lost more rapidly than the methyl group, which indicates that it is being metabolized via the glyoxylate cycle rather than the TCA cycle (Fig. 7.16).

When yeast is grown aerobically on glucose during the sporulation phase, the optimum time for transfer occurs at 18 h. Before this time, the level of alcohol produced by fermentation increases, but after this

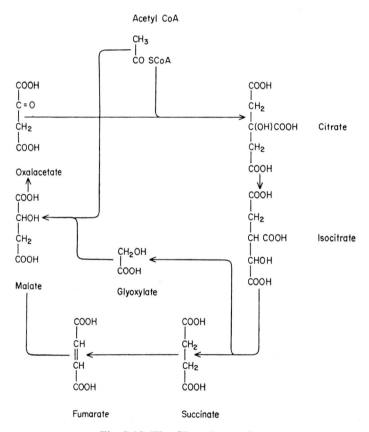

Fig. 7.16. The Glyoxylate cycle.

time, the level decreases as the alcohol is converted to acetate. Since most yeast strains cannot metabolize acetate in the log phase, Croes (1967b) considered that the conversion of ethanol to acetate and the development of the capacity to metabolize acetate was critical. During the early stages of sporulation, the level of amino acids is depressed. Croes postulated that the glyoxylate cycle, which is essential for the production of the carbon skeletons of amino acids, was not very active in the early stages of sporulation and that this resulted in the depletion of the amino acid pool and the subsequent levelling off in the protein content of the cells. The more direct evidence from studies on isocitrate lyase and isotope studies using ^{14}C labelled acetate suggest that this hypothesis is not correct. The level of the glyoxylate cycle does, however, appear to be critical in determining the transition from vegetative growth and mitosis to sporulation and meiosis. Addition of glyoxylate to the acetate sporulation medium results in a continuation of mitosis for up to 24 h after transfer, but rather than inhibiting ascus formation as might be expected, it actually accelerates the rate of formation of asci (Bettelheim and Gay, 1963). Glyoxylate also replaces the requirement for carbon dioxide in the sporulation phase.

Although there are two enzyme systems in yeasts which are potentially capable of synthesizing yeast glycogen, the glycogen synthetase and the glycogen phosphorylase pathways (Fig. 7.17), the glycogen synthetase appears to be the most important in glycogen synthesis. This is not surprising since the equilibrium constant for this is 250 compared with 3 for glycogen synthesis using glycogen phosphorylase. Two glycogen synthetase enzymes have been isolated, one which requires glucose, 6-phosphate (D), and the other which acts independently of glucose, 6-phosphate (I). Similar enzymes in mammals are interconvertible (Cohen, 1968). The increase in the rate of glycogen synthesis at the onset of yeast sporulation is associated with an increase in the level of the I enzyme. Mutants which lack this enzyme do not form glycogen under these conditions. Synthesis of glycogen from glucose, 1-phosphate using the glycogen phosphorylase enzyme, is probably further controlled by the levels of UDPG and glucose, 6-phosphate which inhibit the binding of glucose, 1-phosphate to the phosphorylase. These could be expected to occur at high levels in conditions which favour glycogen synthesis (Rothman and Cabib, 1970).

Glycogen breakdown during endogenous respiration is controlled by glycogen phosphorylase. This enzyme exists both as a dimer (a) and a tetramer (b) in the yeast cells. The tetramer can be converted to the dimer by the action of an ATP dependent protein kinase, and the reverse reaction is catalysed by protein phosphatases. Unlike mammalian

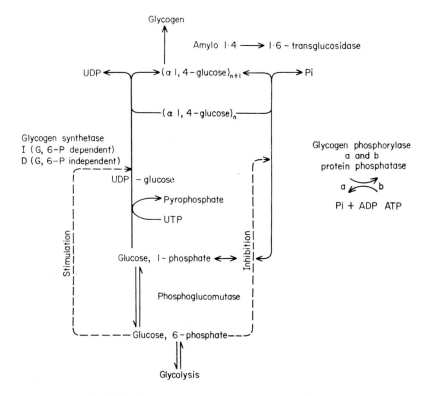

Fig. 7.17. Glycogen metabolism in sporulating yeast.

glycogen phosphorylase, both forms are active and both are inhibited by glucose, 6-phosphate (Fossett *et al.*, 1971). The high level of glucose, 6-phosphate found during periods of glycogen synthesis should limit glycogen breakdown at this time.

In any study of differentiation, it is of considerable interest to know whether the informational macromolecules, DNA, RNA and protein, are being synthesized. Although the levels of RNA and protein decrease during sporulatin, studies using isotopically labelled precursors indicate that RNA and protein synthesis do occur. Maximal rates of RNA synthesis have been observed at T_{10} and T_{25} and the distribution of labelled precursors indicates that r-RNA and t-RNA are being synthesized (Esposito *et al.*, 1970). m-RNA is not readily identifiable in yeast. Unfortunately, Actinomycin D cannot penetrate the normal yeast cell, so it has not been possible to demonstrate a requirement for m-RNA synthesis, using this inhibitor (Tingle *et al.*, 1973). However, azaguanine, which inhibits the formation of functional messenger also inhibits ascospore formation up to stage V. An important stage in sporulation appears to be the resting stage prior to transfer to the sporulation medium when RNA accumulates and the rate of RNA turnover increases. Evidence has been presented to show that sporulation is more probable in cells containing a high level of RNA and protein at this stage.

Although no exogenous nitrogen is required for sporulation, protein synthesis appears to be essential. Sporulation can be inhibited by the addition of cycloheximide, an inhibitor of extra-mitochondrial protein synthesis in yeast up to the mature ascospore stage. Proteins which give a different antigenic reaction have been observed in ascospores, suggesting a requirement for new proteins. There appear to be two main periods of protein synthesis T_{4-7}, and T_{23-27}; however, again it should be noted that the level of protein and the rate of protein turnover increases in resting cells (Esposito *et al.*, 1969). All *de novo* protein synthesis must involve the reutilization of nitrogenous compounds existing in the cell. The high level of protease activity and low levels of most amino acids indicate that the amino acids required are obtained by the hydrolysis of existing proteins. Small vesicles have been isolated which contain high levels of hydrolytic enzymes and which are more abundant in resting cells than in logarithmically growing cells (Matile *et al.*, 1971).

In attempting to elucidate the critical factors controlling meiosis and sporulation, it is valuable to distinguish between changes which are simply responses to alterations in physiological conditions which would occur in haploid, or non-fertile strains, and those which are peculiar to sporogenesis. The accumulation of glucan and glycogen on acetate medium occurs in non-sporulating cells as well as sporulating cells. Similarly, the rate of incorporation of ^{14}C acetate into lipids and its distribution between different lipid fractions is identical when haploid and diploid cells are grown on acetate. Conversely, the increase in isocitrate lyase observed when fertile diploid cells are transferred to acetate medium does not occur in asporogenic strains. It has also been shown that in $\alpha\alpha$ diploids, premeiotic DNA synthesis is blocked on acetate medium.

Euascomycetes

SEXUAL REPRODUCTION

Morphology and Cytology. The sexual cycle of the Euascomycetes is more complex than that of Hemiascomycetes. Although there is considerable variation between different genera, particularly with respect to structure of the ascocarp, it is possible to outline several different stages which are potentially interesting from a morphogenetic point of view (Fig. 7.18).

Production of antheridia and ascogonia. These usually differ in structure and are frequently produced in close association on the same mycelium. A notable exception to this, however, is the genus *Neurospora* in which

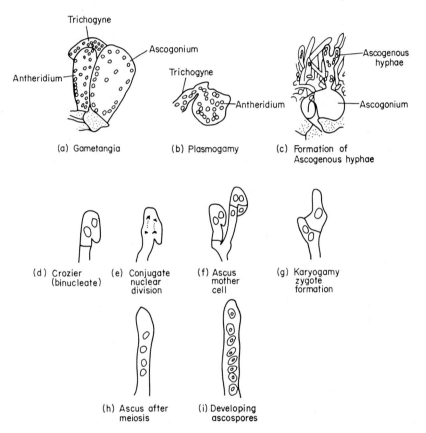

Fig. 7.18. Sexual reproduction in *Pyronema omphalodes*. From Alexopoulos (1962).

no antheridia are produced and special microconidia act as the male gametes. The ascogonium normally produces special hyphae, the trichogynes, which act as the receptors for the male gametes (Fig. 7.18(a)).

Plasmogamy. Nuclei from the male gamete pass through the trichogyne into the ascogonium. In species which produce microconidia or spermatia it is known that trichogynes are able to modify their direction of growth in response to the proximity of compatible male cells. Not all Euascomycetes are heterothallic but those which are have a bipolar system of heterothallism, e.g. *Neurospora crassa* (Fig. 7.18(*b*)).

Ascocarp production. The sexual process in most Euascomycetes occurs in a protective structure known as the ascocarp. This is a resistant structure of pseudoparenchyma or prosenchyma, which is derived from somatic hyphae. The initiation of ascocarp production may occur before the onset of gamete production or may be stimulated by plasmogamy. The ascocarp may completely enclose the ascogonia, e.g. *Aspergillus nidulans,* or may be open at maturity by means of a pore, e.g. *Neurospora crassa,* or may be completely open, e.g. *Ascobolus stercorarius.* These are referred to as cleistothecia, perithecia and apothecia respectively (Fig. 7.19). Within the mature ascocarp, the mature asci may be arranged more or less at random or in a single layer, the hymenium.

A further complication is added by the relationship of the perithecium to the remainder of the mycelium, whether it is simply on the surface of the mycelium or embedded in a compact mat of hyphae, the stroma.

Production of ascogenous hyphae. Nuclear fusion does not follow immediately after plasmogamy in this group. Initially special ascogenous hyphae, which are in the first instance multinucleate, grow out from the ascogonium. Septa then form in a manner which separates off a uninucleate cell at the tip of the hyphae then a series of binucleate cells in the rest of the hyphae. Since nuclear fusion occurs in one of these binucleate cells, it can be assumed that one is derived from the ascogonium and the other from the male gamete if recombination between the two genomes is to occur. The mechanism by which this dikaryon is produced and its relationship to dikaryon production in Basidiomycetes is not known (Fig. 7.18(*c*)).

Nuclear fusion. The cell in which nuclear fusion is to occur can be easily recognized since it becomes crozier shaped. Prior to diploidization, the two nuclei divide in the same plane in such a way as to leave one of each daughter nucleus in the tip of the cell which is then cut off by the formation of two septa to produce a binucleate ascus mother cell (Fig. 7.18 (*d-g*)).

Meiosis and ascus formation. Meiosis occurs immediately after the formation of the single diploid nucleus and is normally followed by a single mitotic division producing eight nuclei which become separated into the eight developing ascospores. The mechanism by which the nuclei remain precisely ordered during meiosis and the subsequent mitosis in *Sordaria* and *Neurospora* presents a fascinating problem.

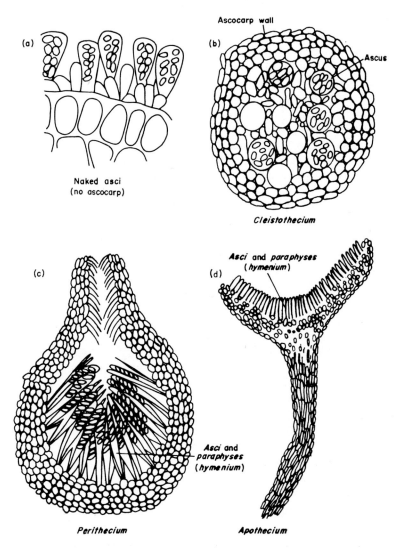

Fig. 7.19. Ascocarp structure in the Euascomycetes. From Alexopoulos (1962).

Physiology and Biochemistry. Since most of the complex events outlined occur within the ascocarp, it is difficult to carry out cytochemical studies and almost impossible to employ biochemical techniques. As a result, our knowledge of the detailed control of sexual reproduction in the Euascomycetes is limited. Some information is however available on the control of the initiation of ascocarp production.

In those species which produce antheridia, these are developed in similar physiological conditions to the ascogonia. Unfortunately, no information is available indicating the critical biochemical events which determine whether a hypha develops into an antheridium or an ascogonium. In species such as *Neurospora crassa*, in which conidia function as the male gametes, the conidia are produced in conditions which favour asexual reproduction rather than those favouring sexual reproduction and are best discussed in this context (Chapter 6). Although media which favour sexual reproduction are known, conflicting results have been obtained in studies aimed at determining which nutrients are critical in such media. It seems, however, that the carbon to nitrogen ratio in the medium is important. In *Neurospora*, nitrate exhaustion is essential for the development of protoperithecia. Stimulation of perithecium formation by sugar phosphates, especially phosphoglyceric acid, has been reported in *Chaetomium globosum*, and by thiamine in *Chaetomium cochloides* and *Chaetomium convolutum*. An extract of jute which stimulated perithecium formation in *Chaetomium globosum* was found to be replaceable by Ca^{++} ions, and in *Sordaria* a medium containing biotin was found to be necessary for perithecium formation (Turian, 1966).

There have been several reports of a requirement for light for the induction of sexual reproduction in the Euascomycetes. Irradiation with U.V. light (290 nm) has been shown to induce perithecium formation in *Pleospora herbarum*. The optimum temperature for induction is between 24°C and 27°C, but the perithecia induced at this temperature do not mature unless the temperature is then reduced to 5–10°C for 21 days (Leach, 1971). In *Pyronema confluens*, light has been shown to be essential for perithecium formation and for the production of certain orange carotenoid pigments in both the mycelium and the apothecia. These two processes do not, however, appear to be closely related since it has been demonstrated that exposure to U.V. light in the region of 300–400 nm is essential for apothecium formation, although the orange pigments do not develop unless the mycelium is exposed to light of 360–580 nm wavelength. Furthermore, the colourless apothecia produced by treatment with U.V. light did not contain any colourless polyenes. It is apparent that two photoreceptors are involved, one in polyene synthesis and the other in apothecium formation (Carlile and Friend, 1956).

Conclusive evidence of sex hormones in the Euascomycetes is still lacking, although it has been reported that the trichogyne can alter its direction of growth towards oidia in *Ascobolus stercorarius* in a manner resembling a chemotactic response (Bistis, 1957). Sterols have been implicated in the control of perithecium formation in *Sordaria fimicola*

(Elliot, 1969). Growth and perithecium formation were found to be inhibited by four hypocholesteremic drugs. The inhibition of growth was reversed by the addition of oleic acid but not cholesterol to the medium suggesting that the action of the drugs on growth was not directly concerned with steroid synthesis. The inhibition of perithecium production was not reversed by either cholesterol or oleic acid alone. However after treatment with one of the drugs, addition of the two together did reverse the inhibition. This drug appears to act at a different site in growth and perithecium production. It seems probable that alterations in the steroid metabolism of the cell are involved in the initiation of perithecium formation. When sub-inhibitory concentrations of the drugs were used, a biphasic growth pattern was observed. At the onset of the second of these phases the reduced growth rate was accompanied by the formation of a ring of perithecia at the hyphal front. Further rings of perithecia were developed at intervals during this phase and the total number formed was twice as high as the number formed in the drug free controls.

Attempts to correlate the development of the ascocarp with specific biochemical events have mainly centred around melanin formation and the two enzymes associated with it, tyrosinase and laccase. Laccase formation is correlated with perithecium development in *Podospora anserina* (Bu'Lock, 1961) and mutants which affect both melanin formation and ascocarp development have been isolated. These mutants have a qualitatively altered laccase. However, melanin production and perithecium development can readily be separated on different growth media. *Podospora anserina* grown on a glucose or fructose medium produces melanin but no mature perithecia, whereas pigment production is poor and perithecium development abundant on a saccharose medium which is a poor carbon substrate (Turian, 1966).

In *Neurospora crassa*, melanin production is also associated with perithecial production and the level of tyrosinase increases during the sexual phase. This correlation between tyrosinase and perithecium development has also been observed in *Glomerella cingulata* and in *Hypomyces solani* (Esser and Kuenen, 1967; Wilson and Baker, 1969). In *Hypomyces solani*, the normal period of cold required for the induction of perithecium development can be replaced by treatment of the vegetative mycelium with tyrosinase. In *Neurospora crassa*, growth at 35°C on minimal medium inhibits melanin and perithecium production (Westergaard and Mitchell, 1947) and similar results can be obtained by adding either a high level of nitrogen or an inhibitor of tyrosinase to the medium (Hirsch, 1954). Structural genes and repressor genes which are concerned with the inhibition of both tyrosinase production and perithecium development have been detected. Female sterile mutants are known however in which tyrosinase can be induced by the addition of aromatic amino acids which have no effect on perithecium production. Thus the two functions can be separated genetically. Tyrosinase is not essential for growth, being mainly concerned with the oxidation of mono- and dihydroxyphenols. It appears to be produced in conditions which are unfavourable for growth, and protein synthesis and its production may be dependent upon the rate of breakdown of proteins and release of aromatic amino acids. The level of sulphate or substrates derived from sulphate is also critical in determining the level of tyrosinase, high levels repressing its synthesis (Esser and Kuenen, 1967).

Ascogonia in both *Penicillium* and *Aspergillus* spp. take up basophilic dyes which normally indicate the presence of high levels of nucleic

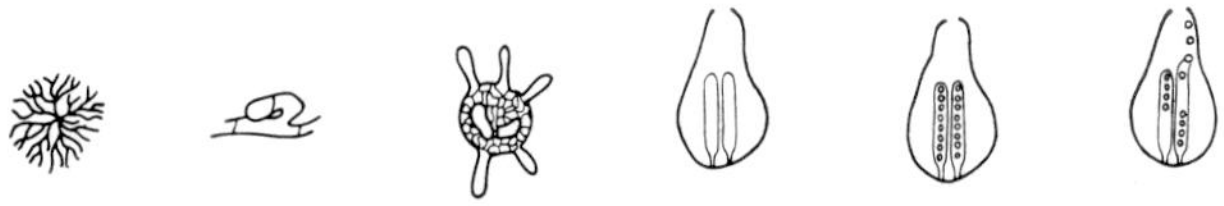

Species	No reproductive organs	Ascogonia but no protoperithecia	Protoperithecia but no perithecia	Sterile perithecia	Fertile perithecia but no spore discharge
Sordaria macrospora	c r	cit spd p	pl f l	s min pa	n in m
Sordaria finicola	St-2	–	St-l	a^3	St-5
Glomerella cingulata	A^1	F^1 St-l	–	B^2 dw l	–
Podospora anserina	i	pa sp			
		al fu	i lg	–	–

Fig. 7.20. Isolation of mutants affecting sexual morphogenes is in the Euascomycetes. From Esser and Kuenen (1967).

acid. In *Neurospora*, the ratio of RNA to DNA is higher in protoperithecial mycelia than in a conidial mycelium (Turian, 1966). Since it is unlikely that the small amount of tissue in the perithecia would affect these values, this suggests that biochemical changes occur in the whole mycelium during sexual reproduction.

Although *Neurospora sitophila* can develop perithecia at oxygen levels as low as 1%, the observation that "poky" mutants cannot act as the female parent in crosses indicates that oxidative metabolism is probably essential for perithecium formation.

Genetic analysis of the morphogenesis of fruiting bodies in the Euascomycetes has been facilitated by the existence of synthetic media which permit sexual reproduction. Different mutations affecting perithecial development and the morphogenetic stage at which they act are shown in Fig. 7.20. The existence of mutants which are only inherited from the ascogonial parent indicates the role of cytoplasmic inheritance in the control of morphogenesis. With reference to this, it has been observed that a loss of the capacity for perithecium formation occurs after prolonged propagation by conidia in *Aspergillus glaucus*. However, the sexual capacity was restored immediately by propagation from the few ascospores which were produced by these less fertile strains. This appears to be a clear-cut example of the role played by sexual reproduction in "rejuvenating" the cytoplasm and restoring it to a basic competence at the beginning of each sexual cycle (Mather and Jinks, 1958). This phenomenon has been well documented in protozoa such as *Paramecium*.

7.4   Sexual Reproduction in the Basidiomycetes

Introduction

The Basidiomycetes are characterized by the production of basidiospores on the outside of the spore-producing structure, the basidium. There are two subclasses of the Basidiomycetes, the Homobasidiomycetes and the Heterobasidiomycetes. The Heterobasidiomycetes include the jelly fungi, the rusts and the smuts in which the basidium is divided into two zones; the hypobasidium and the epibasidium, a tubular structure derived from the hypobasidium which produces the basidiospores. Unfortunately, little is known of the control of sexual reproduction in this group. The rusts and smuts have very complex life cycles and their parasitic mode of nutrition presents difficulties for experimental studies. However, the recent successful axenic culture of rust fungi may overcome some of these difficulties. It is possible that some insight into their development could be obtained by studying the basidiomycetous yeasts, e.g. *Sporobolomyces*, which are thought to be closely related to the smuts.

The Homobasidiomycetes, which produce basidiospores directly on the basidium, have proved much more amenable to both genetic and physiological studies. The most studied species, *Coprinus lagopus*, *Schizophyllum commune* and the commercial mushroom, *Agaricus bisporus*, are gill fungi or mushrooms which produce their basidiospores on the surface of gills which develop on a conspicuous basidiocarp.

The life cycle of the gill fungi is relatively complex (Fig. 7.21). The haploid, uninucleate basidiospores germinate in the substrate and produce an extended, haploid uninucleate mycelium which constitutes the vegetative phase of the life cycle. Asexual reproduction is not well developed and is by fragmentation and oidium formation, e.g. *Coprinus lagopus*. The onset of the sexual phase is marked by the fusion of two compatible strains to form a dikaryon. After plasmogamy, reciprocal

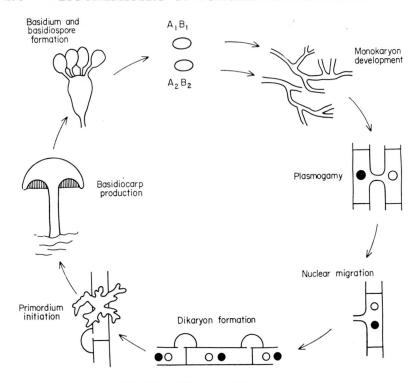

Fig. 7.21. Life cycle of the Agaricales.

exchanges of nuclei occur and a heterokaryon is formed by the migration of nuclei throughout the existing monokaryotic mycelia, giving
rise to a heterokaryon. The dikaryon is developed by subsequent
growth around the periphery of the newly formed heterokaryon. The
dikaryon may continue to produce an extensive mycelium in the substrate before sporophore primordia are initiated. Once initiated, these
grow rapidly to produce the normal mushroom type of basidiocarp.
The whole of the basidiocarp is produced from dikaryotic hyphae. The
final stages in the sexual cycle occur in the surface layers of the gills
which in the Agaricales are suspended from the underside of the pileus.
Hyphae in the surface layers of the gills, the hymenium, develop
specialized terminal cells, the basidia. The basidium, which is initially
binucleate, is the site of nuclear fusion and meiosis so the diploid phase
is a very short period during the maturation of the basidium. Four
basidiospores normally arise from sterigmata on the exposed, outer
surface of the basidium, and into these pass the four products of meiosis.
The mature basidiospores are released by an explosive mechanism.

Sexual morphogenesis can be considered in three stages, plasmogamy and dikaryon formation, the initiation of the basidiocarp primordium and the development and maturation of the basidiocarp.

Cytology and Genetics of Dikaryon Formation

Dikaryon formation in *Schizophyllum commune* and most other basidiomycetes is controlled by two multi-allelic genes, A and B. Successful dikaryon formation requires that the nuclei of the two homokaryons involved in plasmogamy contain different alleles at these two loci. In compatible matings a special form of cell division is initiated involving the production of clamp connections and ensuring that the new mycelium is composed of binucleate cells in which one nucleus of each mating type is present. This requirement is essential if genetic recombination is to occur at a later stage in the life cycle. The sequence of events ocurring in such a cell division is shown in Fig. 7.22.

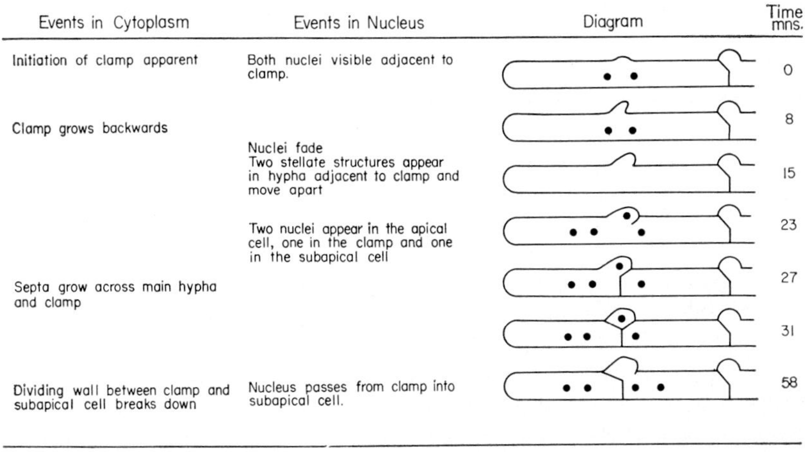

Events in Cytoplasm	Events in Nucleus	Diagram	Time mns.
Initiation of clamp apparent	Both nuclei visible adjacent to clamp.		0
Clamp grows backwards			8
	Nuclei fade. Two stellate structures appear in hypha adjacent to clamp and move apart		15
	Two nuclei appear in the apical cell, one in the clamp and one in the subapical cell		23
Septa grow across main hypha and clamp			27
			31
Dividing wall between clamp and subapical cell breaks down	Nucleus passes from clamp into subapical cell.		58

Fig. 7.22. Nuclear division in a dikaryon of *Schizophyllum commune*. After Niederpruem and Jersild (1972).

Plasmogamy can occur between non-compatible strains, suggesting that it is not controlled by the incompatibility factors. However, a true dikaryon is not established. The results of such incompatible crosses have led to a limited understanding of the role of the incompatibility factors (Table 7.5). In a common B heterokaryon, it is probable that the B locus does not function normally and that the development which does occur is controlled by the A locus. This is referred to as the "A" morphogenetic sequence. A "B" morphogenetic sequence can be

TABLE 7.5

Cross	Description	Fertility	Morphological characterization of heterokaryon
$A_1B_1 \times A_2B_2$	Compatible	Fertile, fruit bodies formed	(a) Nuclear migration (b) True clamp connections produced
$A_1B_1 \times A_1B_2$	Common A	Sterile	(a) Nuclear migration occurs (b) Common A heterokaryon formed (c) Some simple septa formed
$A_1B_1 \times A_2B_1$	Common B	Sterile	(a) Reduced nuclear migration (b) Limited common B heterokaryon formed exhibiting a barrage appearance (c) Pseudoclamp connections develop (d) Fusion between clamp and adjacent cell does not occur
$A_1B_1 \times A_1B_1$	Common AB	Sterile	(a) No nuclear migration (b) Unstable common AB heterokaryon formed exhibiting barrage

established using a similar argument. In *S. commune*, in common A heterokaryons extensive nuclear migration in the mycelium of both homokaryons occurs. The heterokaryotic mycelium is characterized by a limited development of aerial hyphae and some disruption of the cell

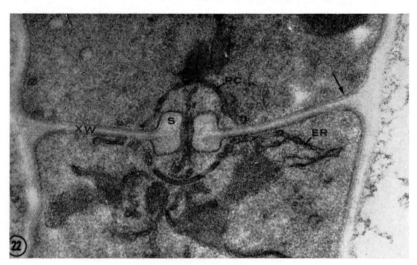

Fig. 7.23 (*a*)

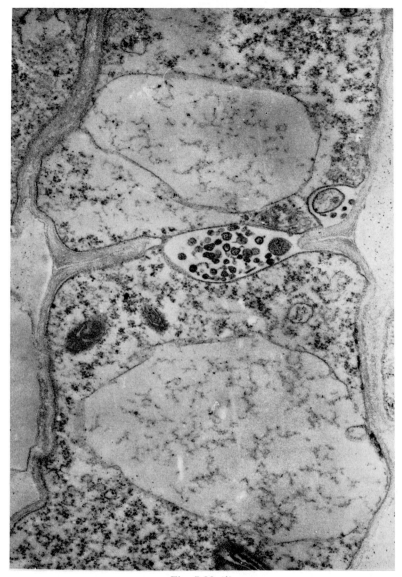

Fig. 7.23 (b)

Fig. 7.23. (a) Structure of dolipore in *Schizophyllum commune*. The cross wall (XW), septal swelling (S), and the continuity of the pore cap (PC) with the endoplasmic reticulum (ER) ×47,000. From Niederpruem and Jersild (1972).

   (b) Invasion of a dolipore by vesicles during dolipore breakdown is a common A heterokaryon of *Coprinus lagopus* ×24,000. From Casselton and Kirkham (personal communication).

septa. The normal septum of the monokaryon is a complex structure known as a dolipore septum. The dolipore septum is characterized by annular thickening around the septal pore. The pore itself is covered by a discontinuous lipoprotein membrane which is referred to as the septal cap. In compatible matings and in common A heterokaryons, dolipore septa are converted to simple septa (Fig. 7.23) allowing nuclear migration through existing hyphae to take place. In the common A heterokaryon in *Coprinus lagopus*, both simple and dolipore septa are visible.

The breakdown of the dolipore septa presents an interesting example of a metabolic process which appears to be precisely localized within the cell since other parts of the cell wall are not affected and septum synthesis and conversion can occur concurrently in the same cell. electron microscopy indicates that vesicles accumulate around the dolipore at the time of breakdown. Although the precise role of these vesicles has not been established, it seems likely that they are involved in carrying hydrolytic enzymes to specific sites in the cell where they are required (see Chap. 5) (Fig. 7.23).

The mechanism by which compatible nuclei migrate during heterokaryon formation is not known. However, recent studies (Casselton and

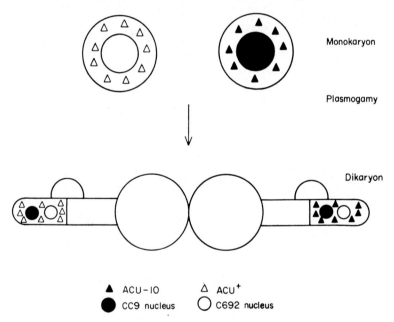

Monokaryon

Plasmogamy

Dikaryon

▲ ACU-10   △ ACU⁺
● CC9 nucleus   ○ C692 nucleus

Fig. 7.24. Diagram illustrating the lack of mitochondrial exchange during plasmogamy in *Coprinus lagopus*. From Casselton and Condit (1972).

Condit, 1972) indicate that the mechanism does not involve proto-plasmic streaming. A mutant *acu* − 10 has been isolated which has an altered cytochrome a_1 spectrum and which is cytoplasmically inherited. Strains carrying *acu* − 10 cannot grow on acetate as the sole carbon source. Compatible matings between this strain and strains which are of a wild type for this marker give rise to a dikaryon in the normal manner with dikaryotic hyphae arising from around the edge of each of the two parent monokaryons. Dikaryotic hyphae were isolated from around each parent and analysed for the presence of the *acu* − 10 marker. It was found that whereas extensive nuclear migration had occurred, no migration of the cytoplasmic markers *acu* − 10 and *acu* + had occurred (Fig. 7.24).

In common B matings, the nuclei only migrate to a limited extent forming a common B heterokaryon which exhibits a characteristic barrage appearance. In the heterokaryotic hyphae the two nuclei

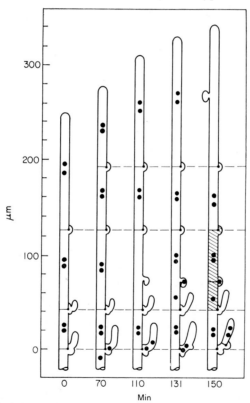

Fig. 7.25. Hyphal branching from clamp connections in dikaryotic hyphae of *Schizophyllum commune*. From Niederpruem and Jersild (1972).

undergo nuclear pairing and conjugated division as in the normal dikaryon and clamp formation is initiated. However, fusion of the clamp with the adjacent cell never occurs so one nucleus remains trapped in the clamp and a true dikaryon never develops.

In addition to these incompatibility factors, a further gene, dik^+, has been shown to be involved in dikaryon formation. Plasmogamy between two compatible strains containing the recessive mutant dik^- gives rise to a diploid mycelium whose morphology resembles that of the homo-karyon. Thus the dik gene appears to be concerned with nuclear fusion.

An intimate relationship between cell division and hyphal branching has been demonstrated recently in a series of elegant studies (Nieder-pruem and Jersild, 1972). The two nuclei in the apical cell move forward during hyphal extension until cell division is initiated. The clamp connection then develops adjacent to the two nuclei which then remain immobile during cell division as hyphal growth continues. The rate of movement of the two nuclei is such that clamp connection formation results in an almost equal division of the parent cell (Fig. 7.25). The formation of a cell wall or septum after mitosis between the two daughter nuclei is a well known phenomenon in higher organisms but has not been previously reported in the fungi.

It has been shown that hyphal branching can occur as a result of new hyphae developing from clamp connections. As a hyphal spur begins to grow from the original clamp connection, the nuclei from the distal cell migrate to a point at the base of the developing daughter

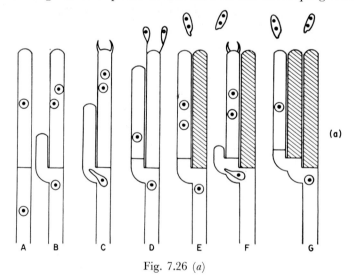

Fig. 7.26 (a)

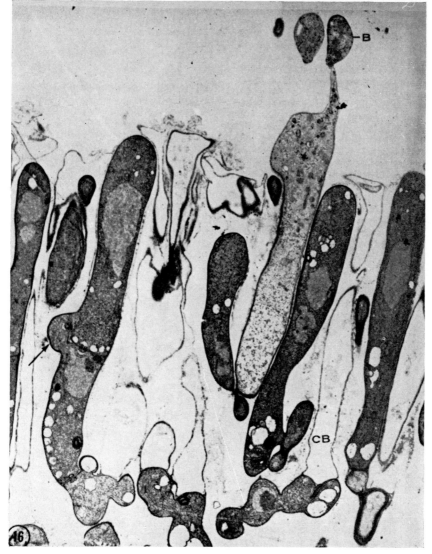

Fig. 7.26. Hymenium development in *Schizophyllum commune*.

(a) Proposed mechanism of basidial proliferation in hymenium of homokaryotic fruit-body of *Schizophyllum commune*. A, uninucleate apical basidium; B, mitosis in mature basidium and subbasidial branch emergence; C, sterigmata emerge on basidium and penultimate nucleus enters probasidium; D, nuclei in apical basidium enter basidiospore initials while mitosis occurs in neck of probasidium and septum delimits neck of outgrowth; E, mitosis occurs in basidiospores and binucleate basidiospores are discharged, leaving collapsed basidium (shaded); mitosis occurs in new basidium; F and G, entire sequence is initiated again. Figure is diagrammatic and not drawn to scale, nor is the time scale known.

(b) Late sporogenesis in hymenium of dihasyotic fruit-body. Note basidiospore (B) attached to sterigma. The clamp connection (arrow) contains complex septa. Cell wall layering is reduced to mound. Collapsed Basidia (CB) are numerous. ×6000. From Niederpruem and Wessels (1969).

hypha where a new cycle of cell division is initiated. This gives rise to an apical, binucleate cell, a new clamp connection adjacent to the original one and the mother hyphal cell which contains the second pair of nuclei produced during the cell division (Fig. 7.25). The elucidation of this mechanism has particular relevance to the mechanism of development of the hymenium in the basidiocarp. Because the first cell division of the daughter hypha always occurs at the base of the mother hypha there is always a clamp connection close to the branch point. Since this can give rise to a new hypha in a similar manner, a whole series of daughter hyphae can arise from the same point on the mother hypha. In this manner a structure resembling the hymenium can be built up (Fig. 7.26). Electron micrographs of the hymenium of *Schizophyllum commune* have shown hyphae grouped together in this manner. At the base of each hypha, structures which appear to be the remains of the clamp connections from which they developed can be observed.

Physiology and Biochemistry of Dikaryon Formation

The mechanism of plasmogamy is difficult to study. Variations in temperature, pH, osmotic pressure and carbohydrate levels appear to have no effect (Ahmed and Miles, 1970). However, it is apparent that heterokaryosis causes a significant alteration in the metabolic activity of the mycelium. In addition to the numerous cytological changes, several biochemical changes have been reported.

Specific enzymes can be identified after electrophoresis on acrylamide gels by using cytochemical stains. With these techniques, it has been possible to recognize changes in isozyme composition between different strains of *S. commune* (Wang and Raper, 1970). The strains used in these experiments were isogenic except for the mating type locus and were produced by back-crossing for 10 or more generations. The differences therefore can be attributed to the effects of the mating type loci. The strains used are shown in Table 7.6.

Fourteen enzymes were studied including NADH and NADPH dehydrogenases. NADP-dependent dehydrogenases, acid phosphatases, esterases and leucine aminopeptidase. Differences were observed in all the enzymes studied except phenol oxidase (Fig. 7.27). These results are surprising for two reasons. Firstly, because there are a large number of enzymes affected by the mating type loci and secondly because phenol oxidase is not affected although it is already known to play a role in sporophore development. It is therefore apparent that dikaryon formation involves a radical metabolic reorganization.

TABLE 7.6. Description of strain used in the experiment described in Fig. 7.27

	Genotype	Mycelial type	Status of sexual morphogenesis
1	A41 B41, A43 B43	Wild type homokaryons	Neither A- nor B-sequence operates
2	A41 B41 + A43 B43	Wild type dikaryon	Both A- and B-sequence operate
3	Amut B41	Mutant-A homokaryon	Only A-sequence operates (mimic of common-B heterokaryon)
4	A43 Bmut	Mutant-B homokaryon	Only B-sequence operates (mimic of common-A heterokaryon)
5	A43 Bmut M11	Modified mutant-B homokaryon	A-sequence inoperative; B-sequence suppressed (mimic of wild-type homokaryon)

In view of the fact that dolipore septa are broken down to simple septa in heterokaryon formation it was of interest to know the relative activities of R glucanase in homokaryons, heterokaryons and dikaryons since R glucanase is known to be involved in the breakdown of cell walls in Basidiomycetes. In the dikaryon $(A_1A_2B_1B_2)$ the level of R glucanase is low in the presence of glucose but increases when glucose is exhausted. The R glucanase level is very low in homokaryons and remains low even when the glucose in the medium is exhausted. The same is true of a common B heterokaryon $(A_1A_2B_1B_1)$. In contrast, the level is increased 30-fold in a common A heterokaryon $(A_1A_1B_1B_2)$ and is independent of glucose concentration. The A and B loci therefore control both the level of the R glucanase and its sensitivity to glucose. Since R glucanase is not produced in the common B heterokaryon, the B locus must be concerned with the presence or absence of the enzyme. Since it is produced at high levels in the common A heterokaryon but is insensitive to glucose concentration, the A locus appears to be concerned with catabolite repression of the enzyme.

In addition, the ratio of S glucan to R glucan is three times higher in a common A heterokaryon than in the normal dikaryon. Studies on isolated cell-wall material indicate that treatment with pronase, lipase or chitinase alone does not dissolve septa. However chitinase treatment followed by R glucanase results in a complete dissolution of the septa while little change can be observed in the cell wall. Thus the high levels of R glucanase in the common A heterokaryon are possibly associated with the conversion of dolipore septa to simple septa.

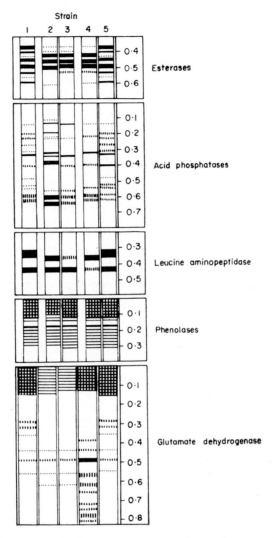

Fig. 7.27. Differences in the isoenzyme content of monokaryons, dikaryons and heterokaryons of *Schizophyllum commune*. From Wang and Raper (1970).

Nuclear migration has been observed in homokaryons of *S. commune* carrying the mutation β_2 at the B locus. These strains exhibit an aberrant hyphal morphology, simple septum formation, high R glucanase levels and a decrease in the level of R glucan in the cell wall. A series of mutations at a locus (M) which is not linked to the mating type have also been isolated which restore the mycelium to the normal

TABLE 7.7. Environmental factors influencing the development of the basidiocarp in some common species of Basidiomycetes. From Taber (1966)

| Fungus | Habitat | High (@ 1%) CO$_2$ | Low humidity | Temperature range (°C) | Geotropic response | Light | | | | pH range |
						Primordium development	Stipe elongation	Pileus expansion	Phototropic response	
Agaricus campestris	Ground	Pileus inhibited	Unfavourable	10–24 (21)	Stipe weak	Inhibitory	Inhibitory	Inhibitory	None	6.9–8
Coprinus lagopus	Ground, dung	—	—	ca. 25	Stipe weak	Inhibited	Base suppressed	Required	Strong	4–9
Coprinus sterquilinus	Dung	—	May be favourable	—	Stipe and gill weak	Inhibited	Base suppressed	Required	Strong	—
Schizophyllum commune	Tree	Unfavourable	Not favourable	22	—	Not required	—	Required	—	—
Collybia velutipes	Tree base	Pileus inhibited	Favourable	10–20 (15)	Strong	Not required	Not required	Required	Strong	5.2–7.2
Polyporus brumalis	Tree	Not inhibitory	Favourable for pileus	15–25 (20)	Strong	Not required	Not required	Required	Strong	4.1
Lentinus lepideus	Tree	—	—	ca. 20	Strong in light	Not required	Not required	Required	Strong	ca. 5.5

monokaryotic state. These mutations not only cause a reduction in the
level of R glucanase in strains carrying the β-2 marker, but also increase
the resistance of the cell walls to degradation by R glucanase in vitro
(Wessels and Koltin, 1972).

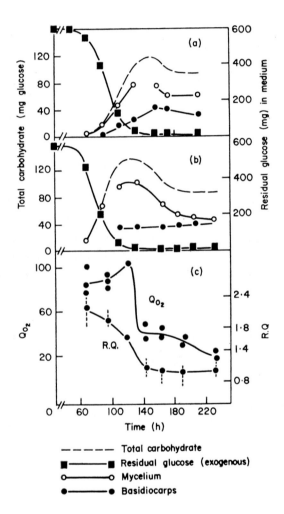

Fig. 7.28. Carbohydrate metabolism during basidiocarp development in *Schizo-phyllum commune*.

  (*a*) values from cup mutant K35
  (*b*) values from normal dikaryon K8
  (*c*) RQ and Q_{O_2} values from strain K8.

From Wessels (1965).

Physiology of Basidiocarp Initiation and Development

The physiological factors controlling primordium initiation and development are very complex and vary from species to species. A summary of the effects of various physical parameters upon different stages of basidiocarp development in some of the more common species is shown in Table 7.7. The development of the basidiocarp in *Agaricus bisporus* is of considerable commercial importance and hence has been studied intensively. This will be discussed later in this section. The nutritional requirements are also complex but in general are in agreement with Klebs' Law which states that reproduction is favoured by conditions which restrict growth.

Physiological studies indicate that both carbohydrate and nitrogen are required for the initiation, whereas only a carbon source is necessary for the growth of primordia. The formation of pilei and the subsequent maturation of the sporophores occurs in the absence of both carbon and nitrogen sources. Since the level of nitrogenous compounds in the mycelium decreases at a time when the level in the sporophore is increasing, it seems possible that nitrogen compounds are transferred from the mycelium to the sporophore. A similar transfer of carbohydrates may occur, although changes in the carbohydrate content of the two fractions do not give any clear indication of this (Fig. 7.28*a* and *b*). Such a result would be anticipated if translocated carbohydrates were metabolized in the sporophore, and therefore did not accumulate.

In *Coprinus lagopus*, it has been shown however, that over the short period of rapid expansion of the basidiocarp between day 12 and 13, the gain in weight of the basidiocarp is exactly equal to the loss in weight of the vegetative mycelium (Table 7.8).

TABLE 7.8. The distribution of dry weight of fungus tissue between mycelium and basidiocarps in a developing plate culture of *Coprinus lagopus*. From Burnett (1968).

Stage of development	Age (days)	Mean dry mycelium (mg)	Weight (4 replicates) basidiocarps (mg)	Total (mg)
Initiation of primordia	8·9	26	0·2	26·2
	9·7	30·1	0	30·1
Enlargement of primordia	10·9	31·6	3·9	35·5
	11·7	*37·2*	*4·5*	*41·7*
Maturation of primordia	12·7	*22·0*	*19·3*	*41·3*
	13·7	30·9	17·5	48·4
	15·0	30·2	29·9	60·1

These results provide a physiological basis for the relationship between the amount of vegetative mycelium and the number of sporophores produced. In *Agaricus bisporus*, the wet weight of basidiocarp produced varies directly with the total dry weight of the culture (Fig. 7.29) suggesting that basidiocarp development is restricted by the amount of reserves in the mycelium.

The period of primordial growth after nitrogen exhaustion is associated with a high respiratory rate, and a respiratory quotient of approximately 2. This continues until the glucose is exhausted, at

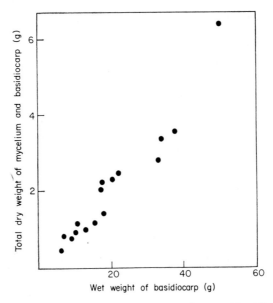

Fig. 7.29. Relationship between the size of the mycelium and basidiocarp production in *Agaricus campestris*. From Burnett (1968).

which time pileus formation is initiated, the rate of respiration drops rapidly, and the R.Q. is reduced to 1. (Fig. 7.28*c*). At 30°C, primordia do not grow, and the period of high respiration does not occur.

There is strong evidence that hormones are involved in the control of basidiocarp formation. Removal of the pileus of the developing sporophore of *Coprinus lagopus* inhibits any further growth of the stipe. Growth can be restored, however, by replacement of the pileus with a few lamellae. Similar results have been reported in *Agaricus bisporus* and *Collybia velutipes*. Removal of the lamellae from half the pileus results in curvature of the stipe away from the side bearing the lamellae. Similar results have been obtained using the classical technique of agar blocks which have been in contact with small fragments of lamellar material (Gruen, 1963, see Fig. 7.30). These studies suggest that the lamellae produce a hormone which travels down the stipe and stimulates growth. The plant hormones, indole acetic acid and gibberellic acid, have both been implicated as the active component. It would not be surprising if these or other "hormones" are involved in controlling the development of such a complex structure as the basidiocarp.

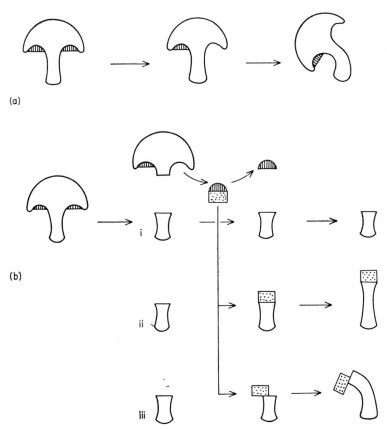

Fig. 7.30. Evidence for hormone action during the development of the basidiocarp in *Agaricus*. From Gruen (1963).
    (*a*) Removal of the lamellae from one side of the pileus results in a curvature in the opposite direction.
    (*b*) Experiments using extracts which have diffused from the lamellae into an agar block.
        (i) Removal of the pileus leads to the inhibition of stipe growth.
        (ii) Application of the agar block permits growth to resume.
        (iii) Unilateral application of the agar block caused growth curvature.

The existence of such hormones would provide a possible biochemical mechanism for the phototropic and geotropic responses observed in the sporophores of the homobasidiomycetes. Hormones from the pileus do not appear to be essential for stipe elongation in *Coprinus cinereus* in which elongation can occur when the stipe is isolated from both the vegetative mycelium and the pileus. During this elongation phase both the level of chitin and the level of chitin synthetase in the stipe increase. Since exogenous nutrients are not provided the synthesis of chitin must be dependent upon the breakdown of other cell constituents not yet identified. The majority of the chitin synthetase activity was found to be associated with the microsomal fraction, but the cell wall fraction did contain up to 30% of the total activity (Gooday, 1973). The association of chitin synthetase with the microsomal fraction suggests that it may be transported from its site of synthesis to the cell wall in vesicles.

Biochemistry of Primordium Initiation and Development

The inhibition of respiration at the onset of pileus formation is not simply the result of carbohydrate depletion but, at least in *Agaricus bisporus*, is caused by the production of a specific inhibitor of respiration (Weaver *et al.*, 1970). The red pigment which accumulates during pileus formation in *A. bisporus* has been demonstrated to be an active inhibitor of respiration. The inhibitor itself is very unstable *in vitro*, but can be generated enzymically from a precursor which is also found in the pileus. Both the precursor and the inhibitor have been characterized and shown to be amides of glutamic acid (Fig. 7.31*a*). The inhibitor is believed to act by blocking sulphydryl enzymes using the mechanism shown in Fig. 7.31*b*.

The decrease in carbohydrate content of the total mycelium after glucose depletion can be attributed, at least in part, to a breakdown in cell-wall polysaccharide. Analysis of cell walls indicates that they contain glucan, chitin, protein and other unidentified compounds (Fig. 7.32). Glucan, which is the major component of the cell wall, can be further subdivided into three fractions based on solubility in alkali. S glucan appears to be predominantly α 1–3 linked. Upon hydrolysis with formic acid, small amounts of xylose are produced in addition to glucose and nigerose (α 1–3 linked glucosyl glucose). Partial hydrolysis of R glucan gives rise to laminaribiose (β 1–3 linked glucosyl glucose) and gentobiose (β 1–6 linked glucosyl glucose). These hydrolysis products suggest that R glucan contains β 1–3 and β 1–6 linkages. The R glucan fraction can be further sub-divided into an R_s fraction, which is soluble, and an R_i fraction, which is insoluble in hot 2N KOH.

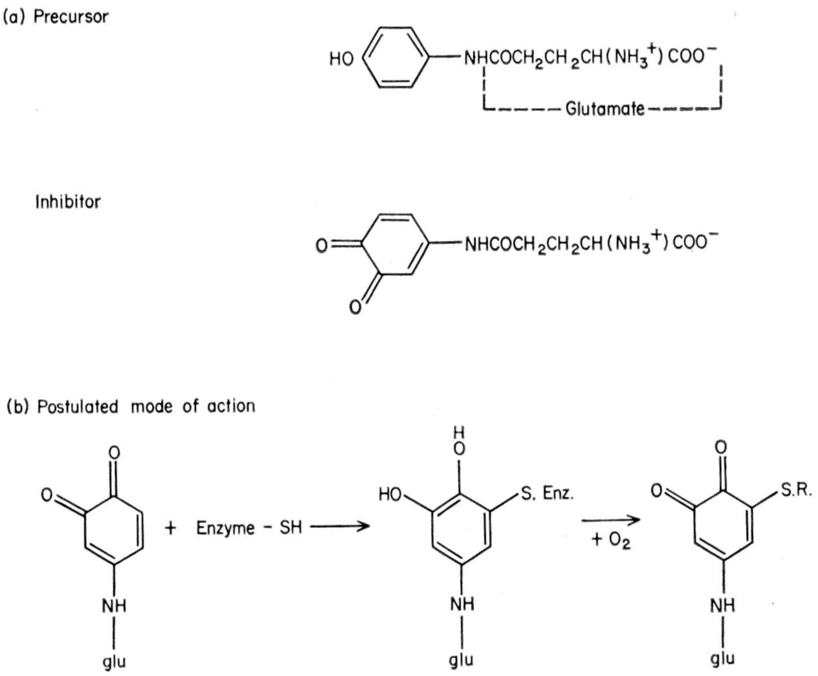

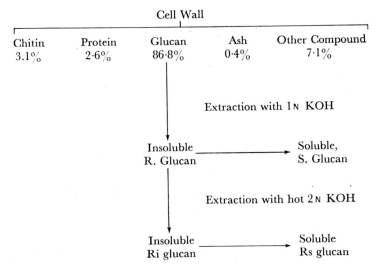

(a) Precursor
HO
NHCOCH2CH2CH(NH3+) COO−
Glutamate
Inhibitor
O
O
NHCOCH2CH2CH(NH3+) COO−
(b) Postulated mode of action
O
O
NH
glu
+ Enzyme - SH
H
O
HO
S. Enz.
NH
glu
+ O2
O
O
S.R.
NH
glu

Fig. 7.31. Structure and mode of action of the respiratory inhibitor isolated from *Agaricus bisporus*. From Weaver *et al.* (1970).

Cell Wall
Chitin
3.1%
Protein
2·6%
Glucan
86·8%
Ash
0·4%
Other Compound
7·1%
Extraction with 1 N KOH
Insoluble
R. Glucan
Soluble,
S. Glucan
Extraction with hot 2 N KOH
Insoluble
Ri glucan
Soluble
Rs glucan

Fig. 7. 32. Fractionation of the cell wall of *Schizophyllum commune*.

The levels of different cell wall fractions varies during the development of fruiting cultures of *Schizophyllum commune*. Comparisons between a wild type strain (K8) and a mutant strain CUP (K35), which exhibits abnormal pileus development, indicate that the decrease in cell wall carbohydrate associated with pileus formation can be attributed primarily to a decrease in the level of R_i glucan. This loss of R_i glucan

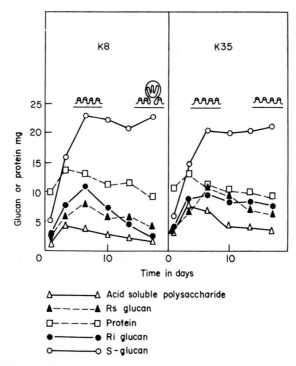

Fig. 7.33. Changes in polysaccharide fractions and protein content during the development of the wild type strain (K8) and the cup mutant (K35) of *Schizophyllum commune*. From Wessels (1965).

can be inhibited by growing the culture at 30°C and so inhibiting pileus formation (Fig. 7.33).

It seems possible that the R_i glucan acts as a reserve source of carbohydrate. During carbon starvation, R_i glucan is broken down in the stroma and in non-developing primordia. An enzyme which breaks down R glucan and which is stimulated by glucose exhaustion has been isolated. The appearance of this enzyme is not, however, restricted to the developing sporophore. It has been found in the *CUP* mutant and in strain *K8* grown at 30°C. Hence the preferential breakdown of R glucan in the developing sporophore cannot be attributed simply to the presence of this enzyme. Experiments on isolated cell-walls show that the enzyme acts only on cell wall material from the wild type strain grown at 20–25°C, and not on 30°C grown material or on cell-wall material from the *CUP* mutant. The difference in morphogenetic response appears to be attributable not to the differential production of the enzyme but to a differential sensitivity of the cell wall which may reflect a subtle difference in structure (Niederpruem and Wessels, 1969).

Other changes in carbohydrate metabolism have been observed in *Coprinus lagopus*. During the period of expansion, differences in the levels of trehalose and glucose in both the stipe and cap can be observed. In the stipe, the level of trehalose increases to 18% of the dry weight and the level of glucose drops from 4% to 2%. In the cap, on the other hand, the level of trehalose increases to 6% then drops to 4% whereas the level of glucose increases from 2% to 7% (Rao and Niederpruem, 1969). The relative levels of two NAD^+ dependent enzymes, sorbitol dehydrogenase and erythritol dehydrogenase are different in the monokaryotic mycelium, the stipe and the cap.

Phenol oxidase activity also appears to be critical for fruit body formation in *Schizophyllum commune*. Its activity in a culture can be detected using the Bavendamm reaction. If 0.1% tannic acid is incorporated into the medium then a dark pigment is produced if phenol oxidase enzymes are present. Using this technique, it was found that mutants lacking phenol oxidase activity were also sterile. The importance of phenol oxidase in fruit-body formation was also indicated by the observation that para-aminobenzoic acid, which has been reported to inhibit pheno oxidase activity also inhibits fruit body formation. Recent results indicate that sterols may also play an important role in sexual morphogenesis in the Basidiomycetes. Changes in the concentrations of sterols, sterol esters and fatty acids have been observed during basidiocarp formation in *Agaricus bisporus*, and in *Coprinus lagopus* the relative abundance of ergosterol, 5-dihydroergosterol and 2,2-dihydroergosterol was found to vary during sexual and asexual reproduction.

Unfortunately, little direct information is available on the importance of protein and RNA synthesis in basidiocarp formation. The ratio of RNA to protein in the mycelium in *Schizophyllum commune* drops at the onset of primordium formation but over the next 24 h the rate of RNA synthesis increases and the ratio of RNA to protein increases again. The significance of these results is not fully understood but it is possible that they indicate an initial breakdown of RNA in the mycelium and a subsequent period of synthesis in the primordia. The accumulation of trichloroacetic acid soluble material which absorbs at 260 nm, indicating low molecular weight polynucleotides at a time when the RNA level is decreasing, may indicate that nucleotides are being translocated from the mycelium and reutilized in the primordia.

Control of Basidiocarp Formation in Agaricus Bisporus

Until recently, it has not been possible to produce sporophores of *Agaricus bisporus* on agar medium. Mushrooms are still produced both in the laboratory and on a commercial scale using a carefully prepared compost. The vegetative mycelium is allowed to grow for two weeks at 23–26°C on a special compost prepared from straw and dung. After this phase the compost is covered with a layer of peat, a process known as "casing". Further growth occurs in the peat until at 2·5 weeks after casing, primordia, which develop into mature mushrooms in 3–4 days, appear.

The role of the casing layer in stimulating the vegetative mycelium to produce primordia has been studied with interest in attempts to obtain a simplified procedure for mushroom production, and several theories to explain this stimulating effect have been put forward. The accumulation in the casing material of carbon dioxide or other volatile compounds such as ethanol, acetone, ethylene oxide, ethyl acetate and acetaldehyde which are known to be produced by mushrooms has been implicated. The existence of other unidentified compounds derived either from the mushroom or from other micro-organisms in the casing material which act as hormones has also been postulated.

It has, however, recently been demonstrated that the casing peat contains a species of bacterium which suppresses vegetative growth in the mushroom and stimulates primordium production. If the peat is sterilized then inoculated with *A. bisporus* the mycelium grows profusely over the peat but no functional primordia are produced. Different groups of micro-organisms were isolated from unsterilized casing material and added singly or in groups to the sterilized material. Using these techniques, several isolates were obtained which stimulated sporophore production. Bacteriological examination of these isolates indicated that they belonged to the species *Pseudomonas putida* or were closely related to it (Hayes *et al.*, 1969). Since this work was completed, an extract has been obtained from *P. putida* which can stimulate sporophore production on defined media. The nature of this extract is still being investigated; however, the results so far indicate that the active component is a chelating agent. Sporophore production can be obtained by growing the vegetative mycelium on malt agar then allowing it to spread out over agar containing either aureomycin, 8,OH-quinoline sulphate or EDTA. These three compounds are all chelating agents. Preliminary results indicate that the critical ion being chelated is iron.

The requirement for microbial interaction in the stimulation of sporophore production may occur in other basidiomycetes which require a complex substrate. The truffle (*Tuber* sp.) only produces fruit bodies when it is growing in a mycorrhizal association with oak. Other mushrooms, such as the padistraw mushroom and the Shii-ta-ké mushroom, fruit on rice straw or branches of trees respectively. However, some 150 spp. of homobasidiomycetes have been grown *in vitro*.

Summary

A complete understanding of the control of sexual reproduction in any organism requires detailed studies on the several phases of the sexual cycle, i.e. gamete formation, plasmogamy, karyogamy, fruit-body formation, meiosis and spore formation.

The fungi provide many experimental systems in which the fundamental biochemical, physiological and genetical problems of the sexual process can be studied. Unfortunately, all the stages referred to cannot be readily studied in the same organism since sexual reproduction frequently involves only a few cells in the mycelium. Furthermore, different stages of sexual development are frequently found on the same mycelium, and in certain fungi critical stages are enclosed within thick-walled protective structures.

The formation of specific reproductive structures, gametangia, which facilitate plasmogamy, occurs in most Phycomycetes and Ascomycetes. However, in the Basidiomycetes and certain Ascomycetes they are absent and plasmogamy occurs between vegetative hyphae. The development of gametangia has been shown to be controlled by hormones in several Phycomycetes, e.g. *Blakeslea trispora* and *Achlya*, although hormones have not as yet been implicated in the formation of oögonia and antheridia in the Ascomycetes. The chemotropic attraction of the trichogyne to microconidia has, however, been reported. Hormones controlling gametangium formation would not be expected to occur in the basidiomycetes, but the possibility that vegetative hyphal fusions may be controlled by hormones should not be overlooked.

The studies carried out on *Hansenula wingei* indicate that plasmogamy is a complex physiological process involving the establishment of a specific cell to cell contact, cell wall synthesis and cell wall breakdown.

A system in which synchronous yeast plasmogamy occurs has recently been described and should be of considerable value for studies on the enzymology of the process (Sena, Radin and Fogel, 1973).

Whereas in the Phycomycetes cell wall breakdown appears to be the critical stage in controlling compatibility, in the Basidiomycetes hyphal fusions occur between incompatible strains but the subsequent events which are essential for dikaryon formation do not occur.

Little data is available on the mechanism and control of karyogamy; however, the isolation of the *dik* mutant which appears to control karyogamy in *Schizophyllum commune* may provide a valuable key to the study of this process.

Although at a morphological and cytological level, fruit body formation in the Ascomycetes and the Basidiomycetes and sclerotium formation exhibit certain similarities, it should not be forgotten that whereas the ascomycete structures are developed from multinucleate homokaryotic or heterokaryotic cells, the basidiocarp develops after plasmogamy from dikaryotic cells, and its production is dependent upon the presence of complementary pairs of incompatibility alleles. Many of the physiological changes observed have been shown to be associated with changes in the dolipore septa. Studies on strains of *S. commune* which permit homokaryotic fruiting may help to establish which biochemical events are specific to fruit-body formation. The paucity of biochemical and genetical data and the diversiy of the data obtained so far militate against making hasty generalizations at this stage. This diversity is clear if stipe extension in *Coprinus lagopus, Coprinus cinereus* and *Agaricus bisporus* is compared. Whereas nutrients from the vegetative mycelium are required for stipe extension in *C. lagopus*, extension can occur in isolated fruit-bodies in the other two species. However, whereas a substance excreted by the lamellae is essential for stipe extension in *A. bisporus*, the stipe of *C. cinereus* elongates at the normal rate when separated from the pileus. Since no nuclear division occurs during this elongation phase, stipe extension in *C. cinereus* provides an ideal system for studying growth as opposed to cell division in a fungus.

The existence of an extended heterokaryotic or dikaryotic phase in the ascomycetes and the Basidiomycetes has provided excellent systems for genetic studies on fungi. In particular, the studies on *Schizophyllum commune* and *C. lagopus* have led to a greater understanding of nuclear migration, nuclear interaction and the control of cell division in the Basidiomycetes.

The development of the sporophore in the Ascomycetes is initiated before plasmogamy, and its development tends to obscure gamete formation, plasmogamy, meiosis and spore formation. Studies on morphogenetic mutants which block different stages of the sexual cycle in *Sordaria macrospora* and other species should prove valuable in elucidating the control of sexual morphogenesis in the higher Ascomycetes.

The relationship between meiosis and mitosis and the mechanism of meiosis present some of the most fascinating problems of biology. Using sporing cultures of *Saccharomyces cerevisiae*, it is possible to study meiosis in a population of cells of which 80% are undergoing a meiotic rather than a mitotic cell cycle. The changes in the levels of enzymes and substrates observed to date indicate that meiosis and spore formation are complex processes; however, it is not possible at the present time to distinguish between events which are essential for meiosis and those which are involved primarily in ascospore formation or are the result of adaptation to the acetate sporulation medium. Genetic studies should again provide a valuable tool for the study of this system. The difficulties of isolating and studying mutants which affect the sexual cycle are well known, although recent studies on yeast sporulation have demonstrated how, with ingenuity, these can be overcome.

In fungi, cell-wall synthesis normally occurs at the hyphal apex alongside existing cell wall material which can act as a template for the spatial orientation of the new material (Chapter 5). Ascospore formation in yeast and other Ascomycetes provides one of the few examples of cell-walls being synthesized in a situation which is isolated from the existing cell-wall. A novel mechanism for controlling the orientation of wall components in the developing cell-wall must exist. The observation in yeast that the ascospore cell-wall is laid down around the nuclear membrane and appears to exclude the ascus cytoplasm provides a mechanism by which the observed 'clearing' of the cytoplasm during sexual reproduction could be brought about. This exclusion of the cytoplasm distinguishes ascopore formation from the formation of asexual spores in which parental cytoplasm is normally included. Further studies on this aspect of yeast sporulation would be of considerable interest.

Further research on sexual processes in fungi should provide a wealth of information much of which will be relevant to all eukaryotic organisms.

Recommended Literature

General Reviews

Alexopoulos, C. J. (1962). "Introductory Mycology". John Wiley and Sons, New York.

Burnett, J. H. (1968). "Fundamentals of Mycology". Edward Arnold, London.

Esser, K. and Kuenen, R. (1967). "Genetics of Fungi". Springer Verlag, Berlin.

Fincham, J. R. S. and Day, P. R. (1971). "Fungal Genetics". (3rd Ed.). Blackwell Scientific, Oxford.

Hendrix, J. W. (1970). Sterols in growth and reproduction of fungi. *Annual Review of Phytopathology* **8,** 111–130.

Machlis, L. (1972). The coming of age of sex hormones in plants. *Mycologia* **64,** 235–247.

Raper, J. R. (1952). Chemical regulation of sexual processes in the *Thallophytes*. *Botanical Reviews* **18,** 447–545.

Raper, J. R. (1966). Life cycles, basic patterns of sexuality and sexual mechanisms. *In* "The Fungi", **2,** 473–511. Ed. Ainsworth, G. C. and Sussman, A. S. Academic Press, London and New York.

Smith, J. E. and Galbraith, J. C. (1971). Biochemical and physiological aspects of differentiation in the fungi. *Advances in Microbial Physiology* **5,** 45–134.

Turian, G. (1969). "Differenciation Fongique". Masson et Cie, Paris.

Differentiation in the Phycomycetes

Austin, D. J., Bu'Lock, J. D. and Drake, D. (1970). The biosynthesis of trisporic acids from β-carotene via retinal and trisporol. *Experientia* **26,** 348–349.

Barksdale, A. W. (1969). Sexual hormones of *Achlya* and other fungi. *Science* **166,** 831–837.

Bu'Lock, J. D., Drake, D. and Winstanley, D. J. (1972). The trisporic acid series of fungal sex hormones: structure specificity and transformations. *Phytochemistry* **11,** 2011–2018.

Burgeff, H. (1924). Untersuchungen über sexualität und parasitisimus bei mucorineen. *Botanische Abhandlungen* **4,** 1–135.

Edwards, J. A., Mills, J. S., Sundeen, J. and Fried, J. H. (1969). The synthesis of the fungal sex hormone antheridiol. *Journal of the American Chemical Society* **91,** 1248–1249.

Elliott, C. G. (1972). Sterols and the production of oospores by *Phytophthora cactorum*. *Journal of General Microbiology* **72,** 321–327.

Gooday, G. W. (1968). The extraction of a sexual hormone from the mycelium of *Mucor mucedo*. *Phytochemistry* **7,** 2103–2105.

Gooday, G. W. (1973). Differentiation in the Mucorales. *Symposium Society General Microbiology* **23,** 269–294.

Grieco, P. A. (1969). The total synthesis of *dl*-sirenin. *Journal of the American Chemical Society* **91,** 5660–5661.

Hutchinson, S. A. (1971). Biological activity of volatile fungal metabolites. *Transactions of the British Mycological Society* **57,** 185–200.

Machlis, L., Williams, M. W. and Rapoport, H. (1966). Production, isolation and characterization of sirenin. *Biochemistry* **5,** 2147–2152.

Shaw, G. (1970). Sporopollenin. *Phytochemical Phylogeny* 31–58. Academic Press, London and New York.

Van den Ende, H. (1968). Relationship between sexuality and carotene synthesis in *Blakeslea trispora*. *Journal of Bacteriology* **96,** 1298–1303.

Van den Ende, H. and Stegwee, D. (1971). Physiology of sex in Mucorales. *Botanical Reviews* **37,** 22–36.

The Yeasts. General Reviews

Fowell, R. R. (1969). Sporulation and hybridisation of yeasts. "The Yeasts". Rose, A. H. and Harrison, J. S. Vol. 1, 303–383. Academic Press, London and New York.

Pfaff, H. J., Miller, M. W. and Mrack, E. M. (1966). "The Life of Yeasts". Harvard University Press,

Tingle, M., Singklar, A. J., Henry, S. A. and Halvorson, H. O. (1973). Ascospore Formation in Yeast. *Symposium for General Microbiology,* **23,** 209–244. Cambridge University Press, London.

Plasmogamy in Yeasts

Conti, S. F. and Brock, T. D. (1965). Electron microscopy of cell fusion in conjugating *Hansenula wingei*. *Journal of Bacteriology* **90,** 524–533.

Crandall, M. A. and Brock, T. D. (1968). Molecular basis of mating in the yeast *Hansenula wingei*. *Bacteriological Reviews* **32,** 139–163.

Duntze, W., Mackay, V. and Manney, T. R. (1970). *Saccharomyces cerevisiae*. A diffusible sex factor. *Science* **168,** 1472–1473.

Herman, A. and Griffin, P. (1967). Regulation of agglutination in *Hansenula wingei*. *Genetics, Princeton* **56,** 564. Abstract.

Sena, E. P., Radin, D. N. and Fogel, S. (1973). Synchronous mating in yeast. *Proceedings of the National Academy of Sciences U.S.A.* **70,** 1373–1377.

Takao, N., Shimoda, C. and Yanagishima, N. (1970). Chemical nature of yeast sexual hormones. *Development, Growth and Differentiation* **12,** 199–205.

Cytology of Yeast Sporulation

Lynn, R. R. and Magee, P. T. (1970). Development of the spore wall during ascospore formation in *Saccharomyces cerevisiae*. *Journal of Cell Biology* **44,** 688–692.

Moens, P. B. (1971). Fine structure of ascospore development in the yeast, *Saccharomyces cerevisiae*. *Canadian Journal of Microbiology* **17,** 507–510.

Moens, P. B. and Rapport, E. (1971). Spindles, spindle plaques and meiosis in the yeast, *Saccharomyces cerevisiae* (*Hansen*). *Journal of Cell Biology* **50,** 344–361.

Rapport, E. (1971). Some fine structure features of meiosis in the yeast *Saccharomyces cerevisiae*. *Canadian Journal of Genetics and Cytology* **13,** 55–62.

Physiology and Biochemistry of Yeast Sporulation

Bettelheim, K. A. and Gay, T. L. (1963). Acetate-glyoxylate medium for sporulation of *Saccharomyces cerevisiae*. *Journal of Applied Bacteriology* **26,** 224–231.

Cohen, G. N. (1968). "Regulation of cell metabolism". Holt Rinehart Winston Inc, New York.

Croes, A. F. (1967*a*). Induction of meiosis in yeast. I. Timing of cytological and biochemical events. *Planta (Berl.)* **76,** 209–226.

Croes, A. F. (1967*b*). Induction of meiosis in yeast. II. Metabolic factors leading to meiosis. *Planta (Berl.)* **76,** 227–237.

Esposito, M. S., Esposito, R. E., Arnaud, M. and Halvorson, H. O. (1969). Acetate utilisation and macromolecular synthesis during sporulation of yeast. *Journal of Bacteriology* **100,** 180–186.

Fosset, M., Muir, L. W., Nielsen, L. D. and Fischer, E. H. (1971). Purification and properties of yeast glycogen phosphorylase a and b. *Biochemistry* **10,** 4105–4113.

Haber, J. E. and Halvorson, H. O. (1972). Cell cycle dependency of sporulation in *Saccharomyces cerevisiae. Journal of Bacteriology* **109,** 1027–1033.

Matile, P., Cortat, M., Wiemken, A. and Frey-Wysling, A. (1971). Isolation of glucanase-containing particles from budding *Saccharomyces cerevisiae. Proceedings of National Academy of Sciences, U.S.A.* **68,** 636–640.

Miyake, S., Sando, N. and Sato, S. (1971). Biochemical changes in yeast during sporulation. II. Acetate metabolism. *Development, Growth and Differentiation* **12,** 285–295.

Watson, D. C. and Berry, D. R. (1973). An improved culture technique for studies on sporulation in *Saccharomyces cerevisiae. Biochemical Society Transactions,* **1,** 1100–1101.

Genetics of Yeast Sporulation

Esposito, M. S. and Esposito, R. E. (1969). The genetic control of sporulation in *Saccharomyces.* I. Isolation of temperature sensitive sporulation deficient mutants. *Genetics, Princeton.* **61,** 79–89.

Esposito, M. S., Esposito, R. E., Arnoud, M. and Halvorson, H. O. (1970). Conditional mutants in meiosis in yeast. *Journal of Bacteriology* **104,** 202–210.

Esposito, R. E., Frink, N., Bernstein, P. and Esposito, M. S. (1972). The genetic control of sporulation in *Saccharomyces.* II. Dominance and complementation of mutants of meiosis and spore formation. *Molecular and General Genetics* **114,** 241–248.

Roth, R. and Fogel, S. (1971). A system selective for yeast mutants deficient in meiotic recombination. *Molecular and General Genetics* **112,** 295–305.

Euascomycetes

Bistis, C. N. (1957). Sexuality in *Ascobolus stercorarius.* Preliminary experiments on various aspects of the sexual process. *American Journal of Botany* **44,** 436–443.

Bu'Lock, J. D. (1961). Intermediary metabolism and antibiotic synthesis. *Advances in Applied Microbiology* **3,** 239–242.

Carlile, M. J. and Friend, J. (1956). Carotenoids and reproduction in *Pyronema confluens. Nature* **178,** 369–370.

Elliot, G. C. (1969). Effects of inhibitors of sterol synthesis on growth of *Sordaria* and *Phytophthora. Journal of General Microbiology* **56,** 331–343.

Hirsch, H. M. (1954). Environmental factors influencing the differentiation of protoperithecia and their relation to tyrosinase and melanin formation in *Neurospora crassa. Physiologia Plantarum* **7,** 72–97.

Horowitz, N. H., Fling, M. McLeod, H. I. and Sueoka, N. (1960). Genetic determination and enzymatic induction of tyrosinase in *Neurospora. Journal of Molecular Biology* **2,** 96–104.

Leach, C. M. (1971). Regulation of perithecium development and maturation in *Pleospora herbarum* by light and temperature. *Transactions of the British Mycological Society* **57,** 295–315.

Mather, K. and Jinks, J. L. (1958). Cytoplasm in sexual reproduction. *Nature* **182,** 1188–1190.

Turian, G. (1966). Morphogenesis in Ascomycetes. *In* "The Fungi". **2,** 339–385. Ed. Ainsworth, G. C. and Sussman, A. S. Academic Press, London and New York.

Westergaard, M. and Mitchell, H. K. (1947). *Neurospora.* V. A synthetic medium favouring sexual reproduction. *American Journal of Botany* **34,** 573–577.

Wilson, D. M. and Baker, R. (1969). Physiology of production of perithecia and microconidia in *Hyphomyces solani* f. sp. *cucurbitae*. *Transactions of the British Mycological Society* **53**, 229–236.

Basidiomycetes. General Reviews

Niederpruem, D. J. and Jersild, R. A. (1972). Cellular aspects of morphogenesis in the mushroom, *Schizophyllum commune*. *Critical Reviews in Microbiology* **1**, 545–576.

Niederpruem, D. J. and Wessels, J. G. (1969). Cytodifferentiation and morphogenesis in *Schizophyllum commune*. *Bacteriological Reviews* **33**, 505–535.

Smith, A. H. (1966). The hyphal structure of the basidiocarp. *In* "The Fungi". **2**, 151–157. Ed. Ainsworth G. C. and Sussman, A. S. Academic Press, London and New York.

Taber, W. A. (1966). Morphogenesis in Basidiomycetes. *In* "The Fungi". **2**, 387–412. Ed. Ainsworth G. C. and Sussman, A. S. Academic Press, London and New York.

Basidiomycetes—Papers

Ahmed, S. S. and Miles, P. G. (1970). Hyphal fusions in *Schizophyllum commune*. 2. Effect of environmental factors. *Mycologia* **62**, 1008–1017.

Casselton, L. A. and Condit, A. (1972). A mitochondrial mutant of *Coprinus lagopus*. *Journal of General Microbiology* **72**, 521–527.

Gooday, G. W. (1973). Activity of chitin synthetase during the development of fruit bodies of the toadstool *Coprinus cinereus*. *Biochemical Society Transactions* **1**, 1105–1107.

Gruen, H. E. (1963). Endogenous growth regulation in carpophores of *Agaricus bisporus*. *Plant Physiology, Lancaster* **38**, 652–666.

Hayes, W. A., Randle, P. E. and Last, F. T. (1969). The nature of the microbial stimulus affecting sporophore formation in *Agaricus bisporus* (Lange) Sing. *Annals of Applied Biology* **64**, 177–187.

Konishi, M. and Hagimoto, H. (1961). Studies on the growth of fruit bodies of fungi. III. Occurrence, formation and destruction of indole acetic acid in the fruit body of *Agaricus bisporus*. (Lange) Sing. *Plant and Cell Physiology* **2**, 425–434.

Leonard, T. J. (1971). Phenoloxidase activity and fruit-body formation in *Schizophyllum commune*. *Journal of Bacteriology* **106**, 162–167.

Rao, P. S. and Niederpruem, D. J. (1969). Carbohydrate metabolism during morphogenesis of *Coprinus lagopus* (sensu Buller). *Journal of Bacteriology* **100**, 1222–1228.

Wang, C. and Raper, J. R. (1970). Isozyme patterns and sexual morphogenesis in *Schizophyllum*. *Proceedings of National Academy of Sciences U.S.A.* **66**, 882–889.

Weaver, R. F., Rajagopalan, K. V., Handler, P., Jeffs, P., Byrne, W. L. and Rosenthal, D. (1970). Isolation of L-glutaminyl, 4-hydroxybenzene and L-glutaminyl 4-benzoquinone II. A natural sulphydryl reagent from sporulating gill tissue of the mushroom *Agaricus bisporus*. *Proceedings of the National Academy of Sciences U.S.A.* **67**, 1050–1056.

Wessels, J. G. H. (1965), Biochemical processes in *Schizophyllum commune*. *Wentia* **13**, 1–113.

Wessels, J. G. H. and Koltin, Y. (1972). R. Glucanase activity and susceptibility of hyphal cell walls to degradation in mutants of *Schizophyllum* with disrupted nuclear migration. *Journal of General Microbiology* **71**, 471–475.

8 Differentiation, Secondary Metabolism and Industrial Mycology*

8.1 Introduction

The use of fungi for the production of commercially important products has increased rapidly over the past half century. The exploitation of fungi by man is not a recent phenomenon. On the contrary, numerous examples are known which indicate that man has been aware of the value of fungi since the dawn of civilization. The fermentation of alcoholic beverages, practised in the days of the Pharaohs, is the best known example of the exploitation of the biochemical activities of a fungus by early man. The use of yeast to leaven bread also dates back to biblical times.

The higher fungi have long been exploited by man. Fruit bodies of basidiomycetes and ascomycetes have been collected and eaten by civilizations throughout the world; the shii-ta-ke mushroom in Japan, the padistraw mushroom in Asia and the field mushrooms in Europe (Hayes and Nair, 1974). The eating of certain fungi for their psychotomimetic effect also became an integral part of religious ceremonies in several primitive societies. The hallucinogenic drugs found in the

* Key to symbols used in this chapter.
X = mass of micro-organisms
t = time
μ = specific growth rate constant
k = exponential growth rate constant
g = generation time
s = substrate concentration
K_s = half saturation constant for limiting substrate
F = feed rate
V = volume
D = dilution rate

basidiomycete *Psilocybe mexicana* are believed to have been used to stimulate religious experience by early civilizations in central America, while *Amanita muscaria* served a similar role in Europe (Wasson, 1971).

The production of alcoholic beverages, biomass (particularly in the form of mushrooms) and the manufacture of therapeutic compounds, together with the production of simple organic compounds, still remain the major fields in which fungi are exploited by man (Smith and Berry, 1974, in press). Apart from the rather unsophisticated techniques used for the production and maintenance of yeast in the brewing and baking industries, the deliberate growth of fungi for commercial ends did not commence until well into the twentieth century. The development of the sulphite process for the production of glycerol by a yeast fermentation, which was widely used during the 1914–1918 war, probably marks the beginning of Industrial Mycology. However, it is since the advent of the submerged culture techniques used in the penicillin fermentation that the greatest expansion in the use of fungi in industry has taken place. At present, increasing numbers of commercially important products are being produced from fungi (Table 8.1) (Smith and Berry, 1974).

The successful production of a fungal metabolite requires a detailed knowledge of the growth characteristics and the physiology of the fungus

TABLE 8.1

Secondary metabolites of economic importance produced by fungi

Secondary metabolite	Organism	Commercial use
Penicillin	*Penicillium chrysogenum*	Antibiotic
	Penicillium notatum	
Cephalosporin	*Cephalosporium acremonium*	Antibiotic
Griseofulvin	*Penicillium patulum*	Antibiotic
Gibberellic acid	*Gibberella fujikuroi*	Plant growth Hormone
Ergot alkaloids	*Claviceps sp.*	Pharmaceutical
Aflatoxin	*Aspergillus flavus*	Pharmaceutical/Research
Kojic acid	*Aspergillus sp.*	Food Flavour
Muscarine	*Clitocybe rivulosa*	Pharmacology

in question. Not only does the production of different metabolites require different physiological conditions but also each fungus is unique in its anatomical, morphological and physiological development. Thus for each fermentation the precise physiological conditions and the correct stage of *development* must be established for maximal product formation. In this chapter the relationship between growth, development and product formation will be discussed.

8.2   Kinetics of fungal growth

Growth in Batch Culture

Growth of all micro-organisms in batch culture can be divided into several different phases. An initial lag phase, in which the organism becomes adapted to the culture conditions, usually precedes a period of unrestricted growth in which the rate of growth is limited by the intrinsic characteristics of the organism and not the availability of nutrients. In batch culture this phase does not continue for long since the limited amount of nutrient present soon becomes exhausted. When this occurs the growth rate decreases during the deceleration phase and finally net growth ceases and the culture enters the stationary phase.

During the period of nutrient excess the growth rate is the maximum possible for that organism in those conditions. In such circumstances most micro-organisms grow exponentially so the increase in microbial mass (X) is proportional to the mass of micro-organisms present and to the specific growth rate (μ).

$$X_t = X_0 e^{\mu t} \tag{1}$$

$$\frac{X_t}{X_0} = e^{\mu t}$$

Taking the logarithm of both sides

$$\log_e \frac{X_t}{X_0} = \mu t$$

$$\log_e X_t - \log_e X_0 = \mu t$$

$$\mu = \frac{\log_e X_t - \log_e X_0}{t} \tag{2}$$

Therefore the value of μ can be obtained from a graph of $\log_e X$ against time using equation (3).

$$\mu = \frac{\log_e X_2 - \log_e X_1}{t_2 - t_1} \tag{3}$$

or converting to common logarithms

$$\mu = \frac{\log_{10} X_2 - \log_{10} X_1}{2 \cdot 3026 \, (t_2 - t_1)} \tag{4}$$

μ is the specific growth rate of the organism and represents the relative increase in mass per unit time. Since in many unicellular micro-organisms the population doubling time is equivalent to the cell generation time, the growth constant is determined using logarithms to base 2 in which case the exponential growth constant (k) obtained represents the number of generations per unit time.

$$\log_2 \frac{X_t}{X_0} = kt$$

which gives (see equation 3)

$$k = \frac{\log_2 X_2 - \log_2 X_1}{(t_2 - t_1)}$$

and

$$k = \frac{\log_{10} X_2 - \log_{10} X_1}{\cdot 301 \, (t_2 - t_1)} \tag{5}$$

since k represents the number of generations per unit time the generation time (g) is equal to the reciprocal of k. The relationship between μ and k is given by equation (6).

$$\mu = k \log_{e2} = 0 \cdot 69k \tag{6}$$

The value of μ obtained during a period of unlimited exponential growth is referred to as μ max. When growth becomes substrate limited the value of μ becomes proportional upon the substrate concentration. The relationship between the specific growth rate and substrate concentration is given by the Monod Equation (7).

$$\frac{dX}{dt} = \mu \, \text{max} \left(\frac{s}{K_s + s} \right) X \tag{7}$$

Continuous Culture

Since the substrates are being metabolized continually throughout a batch culture their concentration also decreases throughout the fermentation. The conditions for growth are in a continuous state of flux. Continuous culture provides a valuable technique by which the concentration of nutrients and the mass of micro-organisms in a fermentation can be kept constant. Continuous culture can be established by feeding nutrients into a culture whose volume (V) is kept constant by the feed rate (F) and the overflow rate. In such a system

Increase in organisms due to growth $\dfrac{dX}{dt} = \mu X$

$$\text{Dilution rate } D = \frac{F}{V}$$

$$\text{Loss of micro-organisms by dilution } \frac{dX}{dt} = -DX$$

Actual change in concentration or mass of micro-organisms

$$\frac{dX}{dt} = \mu X - DX \tag{8}$$

A steady state is established if the increase in mass of micro-organisms due to growth is exactly balanced by the loss due to dilution. At steady state:

$$\mu = D = \frac{F}{V} \tag{9}$$

since $\mu = 0.69k$ (see equation 6).

$$g = \frac{1}{k} = \frac{1}{0.69\mu} = \frac{V}{0.69F} \tag{10}$$

Unlimited growth can be obtained in a turbidostat if the dilution rate is controlled so that the mass of organisms is maintained at a level at which growth is not substrate limited. More frequently however a chemostat is employed in which the dilution rate is controlled at a level which gives a growth rate below μ max and in which one or more substrates is limiting (see equation 7).

Exponential Growth in Filamentous Fungi ?

GENERAL CONSIDERATIONS

In the bacteria each cell divides by fission to give daughter cells which are reproductively identical. In the fungi only the yeasts grow in a similar manner, and even in the yeasts parent and daughter cells can be distinguished morphologically. In the filamentous fungi growth is restricted to the apical region of the filament such that cells produced in previous generations which are now situated in a subapical position do not contribute to growth. A subapical segment of hypha between 100 and 200 μm in length appears to be involved in growth (Righelato, 1974). This is shorter than the length of many fungal cells, e.g. *Schizophyllum commune* at 250 μm (Niederpruem and Jersild, 1972) and *Aspergillus nidulans* at 440–700 μm (Clutterbuck, 1970).

Exponential growth requires that all, or a constant percentage of the mass of the micro-organisms present, contribute to new growth. If all growth takes place in the apical segment of the hyphae then exponential growth will require that new branches are produced at a rate proportional to the rate of increase in cell mass. Whether true exponential growth does occur in filamentous fungi is still a matter for discussion (Righelato, 1974; Mandels, 1965).

It has been reported that the frequency of branching in *Aspergillus nidulaus* was proportional to the specific growth rate when growth on different media was compared (Katz, Goldstein and Rosenberger, 1972). Individual hyphae may also grow exponentially rather than linearly in certain circumstances. Exponential growth has been reported during germ-tube outgrowth in *Anidulaus* and also in the critical period of growth after branch formation. Recently the hypothesis has been presented that fungal growth continues at a rate proportional to the length of the hypha until the rate reaches the maximum characteristic of the organism (Katz, Goldstein and Rosenberger, 1972). The rate of apical growth can be considered to be dependent upon the biosynthetic capacity of the hypha. When this exceeds the capacity of the apical region to utilize the products of biosynthesis a new branch is initiated.

Pellet formation. In static or gently agitated culture, the mycelium does not fragment, so the fungus develops as a loose spherical pellet. Pellet formation is very common during fungal growth in liquid culture, however in more vigorously agitated cultures the pellets are usually small and dense. The growth of such a pellet is eventually restricted by the rate of diffusion of nutrients into the pellet. When this occurs growth becomes restricted to the peripheral zone of the pellet and the growth rate increases in a cubic rather than logarithmic manner. Since it is impossible to maintain homogeneous growth conditions during pellet formation, conditions in which pellet formation does not occur are normally sought. Prevention of pellet formation is ultimately dependent upon the shearing action of the agitation system which breaks up large aggregates of hyphae. However, the sensitivity of the organism to shear varies with the growth rate and hence the degree of branching of the hyphae, and the composition of the medium. A higher frequency of pellet formation has been observed in *Penicillium chrysogenum* at higher growth rates when a high frequency of branching, causing a shorter internode length, would be expected to occur (Righelato, 1974).

Cell ageing and cell death. In *Saccharomyces cerevisiae*, although both the mother and daughter cells produced in a budding cycle continue to divide the mother cell can be clearly distinguished by the presence of the bud scar. Since one bud scar is produced at each cell cycle the age of the yeast mother cell, measured in generations, can be precisely measured by counting the number of bud scars. In a simple batched culture the age distribution of the cells can be predicted and measured experimentally (Table 8.2). It is clear that if all the cells are actually

TABLE 8.2. Age distribution of yeast cells during expoential growth. From Beran, (1969).

No bud scars	0	1	2	3	4	5
% Population	50	25	12.5	6·25	3·125	1·563
No bud scars	6	7	8	9	10	11
% Population	0·781	0·391	0·195	0·098	0·048	0·024

dividing the percentage of cells produced in the last generation and having no bud scars will be 50% and that of all the other age groups will add up to 50% (Beran, 1969).

In filamentous fungi the age of a particular segment of a hypha is not as readily recognizable; however, it is apparent that the age of the hypha increases with increasing distance from the hyphal tip. During pellet growth the actively growing hyphal tips are distributed around the periphery of the pellet and the older hyphae are to be found at the centre of the pellet. During a batch fermentation which has been inoculated with spores the frequency of non-apical, potentially ageing hyphae will rise throughout the fermentation from a value of zero at time zero. The limitation on the supply of nutrients to the centre of the pellet eventually results in the autolysis and death of the older hyphae. If a balance is established between the rate of growth and the rate of autolysis a balanced age distribution may be established later in the fermentation. Whereas in yeast a change in age distribution does not markedly affect the growth rate since all cells are actively dividing, in filamentous fungi in which only young apical cells are dividing, a change in the age distribution would be expected to influence the growth rate.

In continuous culture the situation is somewhat different in that the older cells are continually being removed at a rate which is dependent upon the dilution rate. In this way a balanced age distribution can be established.

8.3   Secondary Metabolism

The Concept

Many fungal products, e.g. penicillin and gibberellic acid, are produced by only a few species of fungi and even then by only a few strains and in small quantities in their natural environment. These compounds, like many others which have been isolated from fungi, do not appear to be essential for the growth of the organism and their production is not associated with the growth phase. Such compounds have been described as secondary metabolites, as opposed to primary metabolites such as amino acids and bases and most lipids and carbohydrates which are essential for growth. The majority of secondary metabolites are produced only by one or a few species and in most cases, the specific role of the compound in the metabolism of the cell is not known. Secondary metabolites have been classified into groups, based on their mode of synthesis in the cell. These are derived from:

(1) acetyl CoA and fatty acids
(2) mevalonic acid
(3) amino acids
(4) sugars
(5) shikimate and/or the aromatic amino acids
(6) intermediates of the tricarboxylic acid cycle
(7) the products of several metabolic pathways

(see also Fig. 8.1).

Several hypotheses have been put forward to explain the production of secondary metabolites in micro-organisms. That they have no function, and are produced as a result of a breakdown in the regulation of the cell's metabolism is considered unlikely in view of the intense pressure of natural selection which can be expected to occur. Those secondary metabolites with antibiotic properties may offer a competitive advantage in nature; however, these only represent a small minority. Two closely related hypotheses have proposed that the specific products of secondary metabolism are not important but the process of secondary metabolism itself is of selective advantage to the organism. It is considered to provide a mechanism by which excess intermediates (Bu'Lock, 1961) or even excess carbohydrates in the medium (Foster, 1949) can be metabolized during adverse growth conditions. Such a mechanism would serve to maintain the cell in a functional state during conditions which prevented growth.

Secondary Metabolism and Growth

The relationship between growth and secondary metabolite formation has been extensively studied (Bu'Lock, 1967, 1974; Demain, 1968, 1972). As a result of studies on the production of gibberellic acid by *Gibberella fujikuroi*, Borrow *et al.* (1961) divided the batched fermentation into five physiological phases which correlated closely with the phases described for a typical batched fermentation based on kinetic data. After the initial lag phase, a balanced phase of growth occurs in which there is a balanced accumulation of nutrients and the chemical composition and morphology of the hyphae remain constant. This phase

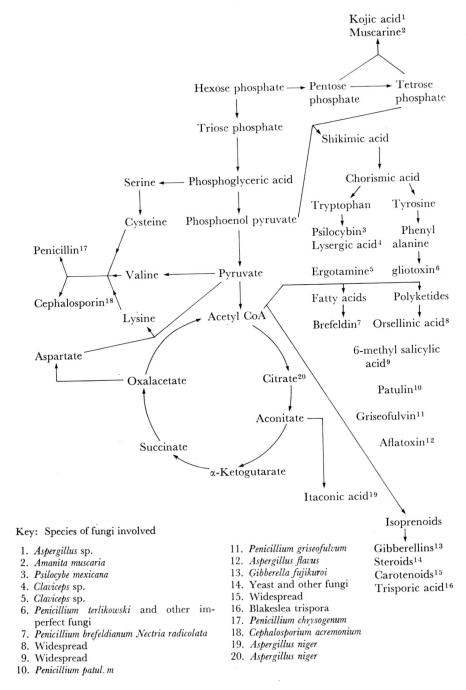

Kojic acid[1]
Muscarine[2]

Hexose phosphate → Pentose — → Tetrose
phosphate   phosphate

Triose phosphate   Shikimic acid

Serine ← Phosphoglyceric acid   Chorismic acid

Cysteine   Phosphoenol pyruvate   Tryptophan   Tyrosine

Penicillin[17]   Psilocybin[3]   Phenyl
                                   Lysergic acid[4] alanine

← Valine ← Pyruvate   Ergotamine[5]   gliotoxin[6]

Cephalosporin[18]   Fatty acids   Polyketides

Lysine   Acetyl CoA   Brefeldin[7]   Orsellinic acid[8]

Aspartate   6-methyl salicylic
                                   acid[9]

Oxalacetate   Citrate[20]   Patulin[10]

Aconitate —   Griseofulvin[11]

Succinate   Aflatoxin[12]

α-Ketogutarate

Itaconic acid[19]

Isoprenoids

Key: Species of fungi involved

1. *Aspergillus* sp.
2. *Amanita muscaria*
3. *Psilocybe mexicana*
4. *Claviceps* sp.
5. *Claviceps* sp.
6. *Penicillium terlikowski* and other im-
   perfect fungi
7. *Penicillium brefeldianum Nectria radicolata*
8. Widespread
9. Widespread
10. *Penicillium patul. m*

11. *Penicillium griseofulvum*
12. *Aspergillus flavus*
13. *Gibberella fujikuroi*
14. Yeast and other fungi
15. Widespread
16. Blakeslea trispora
17. *Penicillium chrysogenum*
18. *Cephalosporium acremonium*
19. *Aspergillus niger*
20. *Aspergillus niger*

Gibberellins[13]
Steroids[14]
Carotenoids[15]
Trisporic acid[16]

Fig. 8.1. Metabolic pathways leading to secondary metabolites. From Turner (1971).

is terminated by the exhaustion of the nitrogen in the medium. During the subsequent storage phase, the dry weight of the mycelium continues to increase due to an accumulation of the storage carbohydrates and fats, while the level of nitrogenous compounds including DNA remains constant. After this storage phase the organism moves into a maintenance phase in which the dry weight remains constant and which continues until the medium and the endogenous reserves of the organism are depleted. Here the terminal phase sets in and autolysis occurs. Secondary metabolites are produced mainly in the storage and maintenance phases (Fig. 8.2). An alternative terminology has been presented by Bu'Lock (Bu'Lock, 1961, 1967). In this, the fermentation is divided

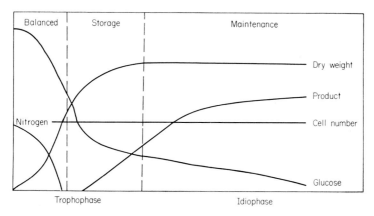

Fig. 8.2. Different phases of fungal fermentations. From Turner (1971).

into an initial trophophase in which the growth rate is high and secondary metabolite production low, and the idiophase in which the growth rate is lower and the rate of secondary metabolite production is increased (Fig. 8.2). Although these terms have a descriptive value, they are neither absolute nor mutually exclusive.

Growth in the unlimited phase is considered to be balanced since the rate of uptake of nutrients into the mycelium and their rate of utilization remains constant, being controlled by the inherent characteristics of the organism rather than the nutrient level. In such growth conditions the whole of the biosynthetic capacity of the cell can be expected to be required to maintain the high rate of growth and cell division. Growth becomes unbalanced when the growth rate is limited by a shortage of one or more nutrients. Those cellular processes which require the limiting substrate become restricted whereas others, which do not require it, are not. Since growth is dependent upon a wide range of biosynthetic activities the absence of any one of several nutrients, carbon, nitrogen, phosphate or trace metals, may restrict it. The remaining non-limiting nutrients may then be diverted into biosyntheses which are not related to growth.

The biosynthesis of secondary metabolites or specific compounds required during differentiation may be stimulated in this manner.

Unlimited growth may occur during the early stages of a batch fermentation or in a turbidostat. In the storage phase of a batch fermentation and in a chemostat, growth is nutrient limited and by definition unbalanced.

It is not possible to define a specific growth rate below which growth can be said to be of the idiophase type. However, such a growth rate can be defined for a specific metabolite produced in the idiophase. Perhaps the only conditions in which true trophophase exists occur when the organism is growing logarithmically, that is when all the biosynthetic capabilities of the cell are geared to growth and cell division. If exponential growth is not achieved, then growth could be considered to be intermediate between trophophase and idiophase (Bu'Lock, 1974).

Since the production of secondary metabolites is not correlated with growth, it is desirable in the industrial production of a secondary metabolite that once the organism has been produced in the fermenter, a minimum of the nutrients supplied are utilized in forming new biomass and a maximum is directed into product formation (Turner, 1971).

These ideas can be illustrated by reference to three examples in which secondary metabolite formation is stimulated by limiting either the carbon, nitrogen or phosphorous feed (Bu'Lock, 1974). In batch cultures of *Claviceps purpurea*, little alkaloid synthesis occurs during the balanced growth phase. The onset of the storage phase is controlled by the depletion of phosphate in the medium. When this occurs growth and protein synthesis are severely limited and a phase of rapid alkaloid synthesis is initiated (Taber, Brar and Giam, 1968).

In the gibberellic acid fermentation the growth rate is controlled by limiting the supply of nitrogen. Chemostat studies using glycine as the nitrogen source have shown that μ max for *Gibberella fujikuroi* is approximately 0.18 h^{-1} and at growth rates approaching this value growth is balanced. At growth rates of 0.10 h^{-1} carbohydrates accumulated but no bikaverin or gibberellin synthesis occurred. Growth rates of less than 0.05 h^{-1} were required for bikaverin formation and of less than 0.01 h^{-1} for gibberellin synthesis (Bu'Lock, 1974).

The use of lactose to reduce the growth rate of *Penicillium chrysogenum* and stimulate penicillin biosynthesis is well established. It has been demonstrated both in extended batch fermentation and in continuous culture that lactose can be replaced by a slow continuous feed of glucose (see p. 300).

Biochemistry of Secondary Metabolite Formation

The details of the biosynthesis of specific secondary metabolites are beyond the scope of this book, even where available. Unfortunately, the biosynthetic pathways leading to many secondary metabolites have not been studied in any detail and there are few examples in which the enzymes involved have been characterized and isolated. It is possible that the production of some secondary metabolites is controlled simply by mass action and the regulation of the activity of existing enzymes. In *Penicillium urticae* and *Penicillium chrysogenum*, glucose catabolism during trophophase occurs primarily via the hexose monophosphate pathway which results in the formation of the large quantities of NADPH required for biosynthesis. In contrast, the glycolytic pathway predominates in the idiophase. The decrease in the rate of biosynthesis of macromolecules associated with the transition from trophophase to idiophase should cause an accumulation of ATP and reduced pyridine nucleotides which will be available for other biosynthetic reactions such as those involved in secondary metabolite production. The biosynthesis of fatty acids, isoprenoid compounds and polyketides from acetate may be further stimulated by increases in the level of the glycolytic intermediates; 3-phosphoglycerate and fructose 1·6-diphosphate which have been shown to stimulate fatty acid synthesis in cell-free extracts of *Saccharomyces cerevisiae*.

The biosynthesis of many secondary metabolites during idiophase does, however, appear to be dependent upon the formation of new enzymes. The biosynthesis of gibberellic acid is inhibited by the addition of inhibitors of RNA and protein synthesis prior to idiophase. Cycloheximide has been shown to inhibit 6-amino salicylic acid formation in *Penicillium urticae* and alkaloid synthesis in *Claviceps* sp. It seems probable that the biosynthesis of secondary metabolites does not simply involve the shunting of primary metabolites down existing but non-active biosynthetic pathways but involves the synthesis of new enzymes and may represent a complete change in the metabolic state of the cell. Such a change is indicated by the alterations in the RQ, Q_{co_2} and the RNA/dry weight ratio often observed during the transition from trophophase to idiophase.

Different regulatory mechanisms which may be operative in the transition from trophophase to idiophase have been discussed by Demain (1972):

(i) Induction of a secondary metabolite by the accumulation of a primary metabolite at the onset of idiophase, e.g. tryptophan acts as an inducer of alkaloid production in *Claviceps* sp.

(ii) Secondary metabolite biosynthesis may be repressed by a primary metabolite during trophophase and derepressed if the level decreases during the idiophase, e.g. penicillin biosynthesis is inhibited by lysine.

(iii) Catabolite repression frequently occurs when high levels of readily utilizable carbohydrates are available during trophophase. This may be the result of the repression of specific enzymes as has been described in *Escherichia coli* or other mechanisms less clearly understood which have been referred to as vegetative suppression (Bu'Lock, 1974), e.g. repression of penicillin production by high concentrations of glucose.

(iv) The importance of phosphate levels suggests that levels of ATP may be critical in certain fermentations, e.g. oosporein synthesis in *Beauveria*.

(v) Secondary metabolites may only be produced when the cell is in a specific developmental state. Such a development may be controlled in fungi by cellular mechanisms which are as yet unknown. In bacteria, sigma factors have been shown to be involved in controlling differentiation. The observed correlation between secondary metabolite production and differentiation lends support to such a mechanism. (See below).

8.4   Differentiation and Secondary Metabolite Formation

As early as 1898, Klebs observed that sporulation was incompatible with growth and made the generalization that reproduction was initiated by factors which check growth (Chapter 6). This theme was developed by Hawker who concluded that conditions permitting spore formation were always more stringent than those permitting growth. Furthermore, the conditions required for sexual reproduction differed more from vegetative growth than those for asexual reproduction.

Frequently it is the availability of nitrogen and/or carbon which is critical. For example it has been shown that conidiation can be induced by either nitrogen limitation as in *Aspergillus niger* or, as in *Penicillium chrysogenum*, by the controlled reduction of the carbohydrate level to a concentration which is high enough to prevent autolysis but too low to promote growth (Chapter 6).

The extensive changes in cellular metabolism associated with these developmental events have been discussed in previous chapters. The similarity between the physiological and biochemical control of secondary metabolism and differentiation suggests that they are closely related cellular processes each of which can benefit from being studied in the context of the other. The onset of both idiophase and differentiation is characterized not only by the production of new metabolites but also by changes in the enzyme composition of the cell. This may simply involve the changes in the level of enzymes which also occur in the vegetative phase, or the synthesis of new enzymes. It is probably incorrect to refer to an enzyme as a secondary metabolite; however, a similar distinction can be made between enzymes which are essential for growth and cell division and those whose production is not correlated with, nor essential for, growth. Many enzymes as yet undiscovered can be assumed to be involved in differentiation and secondary metabolite formation.

The relationship between secondary metabolism and differentiation can be seen most clearly when the production of a secondary metabolite is associated with a particular developmental process. Carotenoid production is often associated with asexual and sexual reproduction. In *Neurospora crassa* carotenoids are produced in the conidia, and in *Blastocladiella* they are produced in the thick-walled sporangia but not in thin-walled sporangia. The differential production of carotenoids in the sporangia of *Blastocladiella* has been correlated with abnormalities in the tricarboxylic acid cycle. In the carotenoid producing cells, the levels of fumarate and malate dehydrogenase were at 50% of normal, and aconitate and -oxoglutarate dehydrogenases were almost inactive. This presents an interesting example of a change in primary metabolism being associated with the production of a specific secondary metabolite, and a specific process of differentiation. In the Mucorales the production of carotenoids is often associated with sexual differentiation (Chapter 7).

The production of polyphenols is also frequently associated with asexual or sexual differentiation. Polyphenol oxidase has not been found in the vegetative mycelium of *Penicillium*, while its synthesis is initiated soon after the induction of sporulation, and its production is correlated with conidiophore and spore formation. The appearance of polyphenols in the mycelium closely follows the appearance of this enzyme. The example of the tyrosinase production during proto-perithecium formation in *Neurospora* has already been discussed (Chapter 7). In the Basidiomycetes the presence of several phenol oxidases can be inferred from the production of a range of phenolic pigments during basidiocarp development in several wood rotting polypores.

8.5   Relevance of Differentiation and Secondary Metabolism to Industrial Mycology

The capacity to control the metabolic state of the cell by adjusting the growth conditions is of fundamental importance in industrial mycology. The principles involved are the same, whether it is a process of differentiation such as asexual or sexual sporulation, or the production of a secondary metabolite which is being controlled.

Specific Fermentation Processes

BASIDIOCARP DEVELOPMENT AND BIOMASS PRODUCTION

In many of the higher Basidiomycetes and Ascomycetes the development of the reproductive structures is the most conspicuous part of the life cycle. It is perhaps not surprising that several of the earliest examples of the exploitation of fungi by man involved the cropping of fruit-bodies of naturally occurring higher fungi followed later by the deliberate cultivation of some of the edible species. At the present time, mushroom cultivation is a major industry accounting for a turnover of £170 million in the United States, and of £40 million in the United Kingdom. The most common species under cultivation is the edible mushroom *Agaricus bisporus*. However, mushroom growing is a world-wide business based on several species of Basidiomycetes in different parts of the world. The The shii-ta-ké mushroom, *Lentinus edodes*, which grows on tree stumps and the padistraw mushroom, *Volvariella volvacea*, which is grown on rice straw, have large commercial markets in Japan and South East Asia respectively. Other edible species such as morells, *Morchella esculenta*, and truffles, *Tuber melanosporum*, have a more limited market (Smith, 1969, 1972; Singer, 1961; Hayes and Nair, 1974).

The cultivation process for each of these can be divided into two phases, the initial vegetative growth phase and the fruiting stage in which nutrients assimilated in the vegetative phase are utilized in the production of the basidiocarp or ascocarp. The delicate flavour associated with these mushrooms is only found in the sporophore and can no doubt be attributed to the production in the sporophore of specific compounds which are not produced in the vegetative mycelium. Many attempts have been made to produce a mycelium in submerged culture, which has the characteristic flavour of the edible mushroom. Success in this would have an immediate commercial advantage, since a large percentage of the mushroom crop is sold as a soup in which the physical characteristics of the mushroom are not recognizable. The production of mushrooms from low-grade carbohydrate and inorganic nitrogen in fermenters would also have a major advantage over other biomass projects in that the product has a flavour which is acceptable to most cultural groups. In other projects a major problem has been the un-acceptability of the product as anything other than an additive for animal feeds. Unfortunately, although the vegetative mycelium grown in liquid culture has a pleasant taste it does not have the characteristic mushroom flavour. The flavour of the fruit-body of *Agaricus bisporus* appears to be particularly elusive. There have been several reports of the production of mycelium with an acceptable mushroom flavour which was subsequently lost on storage, drying or cooking. Even normal sporophores lose their flavour upon drying. There are indications that flavour production is associated with a slow growth rate. Normal strains of *Agaricus blazei* have a definite mushroom flavour, while mutants exhibiting a more rapid growth rate have lost this flavour. Similar results were obtained with *Agaricus campestris*; of 80 strains tested for flavour production, the strain judged to be the best had a slow growth rate (Worgan, 1968).

Morells appear to be more amenable to production in liquid culture. A mycelium in the form of pellets has been produced which retains the typical flavour of the ascocarp.

PRODUCTION OF COMMERCIALLY IMPORTANT SECONDARY METABOLITES BY FRUITING BODIES OF HIGHER FUNGI

Although they cannot be considered to be economically important at the present time, there can be little doubt that production of hallucino-gens by certain species of Basidiomycete was of economic importance in the past in the sense that these fungi were harvested and no doubt traded by primitive tribes. The fruit-bodies of *Psilocybe mexicana* are still consumed in isolated areas of South America, and until relatively recently the consumption of *Amanita muscaria* appears to have been widespread in Europe. Muscarine, which is produced by *Amanita muscaria* and *Clitocybe rivulosa*, affects the parasympathetic nervous system and is a valuable pharmacological compound (Turner, 1971, p. 28). It is also probable that psilocine, psilocybine and the other pharmacologically active compounds in *Psilocybe* will find a use in the treatment of psychiatric diseases.

The hallucinogenic compounds produced by *Psilocybe mexicana* occur only in the sclerotia and sporophore. These two reproductive structures are produced under different environmental conditions from those which promote growth. Sporophore production occurs when the supply of nutrients is severely restricted and is a light dependent process, whereas sclerotium production occurs on a less restricted medium and does not require light.

Lysergic acid (one of a range of alkaloids produced by different species of *Claviceps*) has a small but legal medical market and a larger illegal market as a hallucinogen. Its production has traditionally been a problem of agriculture rather than of fermentation. *Claviceps purpurea*, the most important species, is a parasite of rye. Fields of rye were artificially infected with *Claviceps purpurea*, then after sclerotia had developed in the grain the rye was harvested and the alkaloids extracted

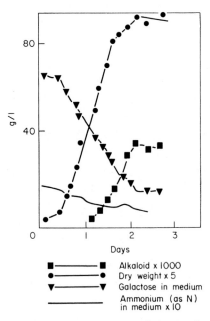

Fig. 8.3. Alkaloid production, mycelial growth and changes in the composition of the medium during the fermentation of *Claviceps purpurea*. From Taber and Vining (1958).

from the sclerotia. The first saprophytic production of ergot alkaloids in a fermenter occurred in 1955 (Vining and Taber, 1963, Taber, 1967). Although in the natural environment the alkaloids are only produced during the development of the sclerotium, in the saprophytic method of production they are produced in the absence of this structure. The conditions required for alkaloid production in the fermenter do, however, differ from those optimal for vegetative growth, and growth suppressants such as phenyl acetic acid and furfural stimulate alkaloid production. It is therefore possible that the mycelium is in a physiologically but not a morphologically differentiated state (Fig. 8.3).

There is an interesting reference to a fungus "Chagi" growing on trees in Siberia which was reputed to cure cancer, in Solzhenitzyn's "Cancer Ward". In "Essays in Biosynthesis and Microbial Development", Bu'Lock refers to a fungal lignin based on a hispidin monomer which occurs in *Polyporus hispidus*. The monomer tends to accumulate at certain stages in the development of the sporophore. Compounds which resemble hispidin have been isolated from *Poria obliqua* which grows on birch trees in Siberia and Scotland. Russian workers claim that this fungus has medicinal properties, and it may be the "Chagi" referred to by Solzhenitzyn.

Vegetative Differentiation and Product Formation

Cephalosporin production. The biosynthesis of cephalosporin provides a very clear-cut example of the role of differentiation in controlling product formation. *Cephalosporin C*, a peptide antibiotic closely related to penicillin, is produced by the Hyphomycete, *Cephalosporium acremonium*. It has been shown, using isotope labelling techniques, that the β-lactam ring of cephalosporin is derived from cysteine and valine and that the α-aminoadipic acid side chain is derived from the same biosynthetic pathway as lysine. *C. acremonium* has a rather complex vegetative growth cycle, in which four distinct types of structure have been described; normal hyphae, arthrospores, conidia and germlings. During the early stages of the cephalosporin fermentation, the hyphal form predominates and only 30% of the mycelium is present as arthrospores. At the end of the growth phase, the level of arthrospores has increased to around 60%. The period of rapid cephalosporin synthesis coincides with this phase. The correlation between cephalosporin synthesis and arthrospore formation was confirmed by separating out the different mycelial forms on a sucrose gradient and testing their ability to incorporate isotopically labelled L-valine into cephalosporin. The majority of the biosynthetic activity was found to be associated with the arthrospore fraction (Fig. 8.4). Genetic studies have provided further evidence of this relationship. A direct linear relationship was observed between the percentage arthrospore production and cephalosporin production in a series of mutants isolated from a strain selection programme. The addition of methionine to the fermentation stimulates both cephalosporin synthesis and arthrospore formation (Fig. 8.5) (Nash and Huber, 1971).

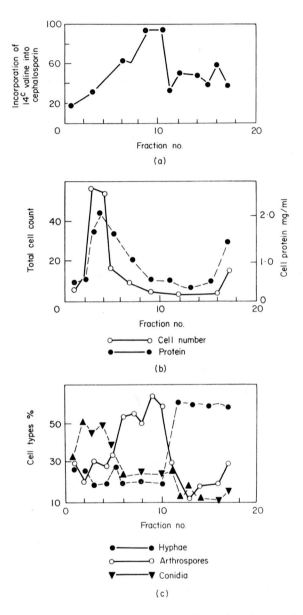

Fig. 8.4. Relationship between arthrospore formation and cephalosporin synthesis in *Cephalosporium acremonium*. From Nash and Huber (1971).

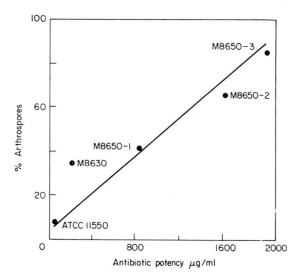

Fig. 8.5. Relationship between arthrospore development and antibiotic production in different strains of *Cephalosporium acremonium*. From Nash and Huber (1971).

Product Formation During Idiophase

Penicillin production. Some of the earliest studies on the penicillin fermentation have suggested that *Penicillium chrysogenum* exists in several different morphological states. The growth phase is characterized by long, thin, relatively unbranched hyphae compared with the penicillin-producing stage in which the hyphae are shorter, thicker and more branched. However, these different forms are not as distinctive as is the formation of arthrospores in the cephalosporin fermentation (Hockenhull, 1963). Early studies suggested that the period of maximal penicillin production was associated with a reduced growth rate. When glucose was used as a carbon source, the rate of growth was limited by the slower rate of metabolism of lactose, and the rate of penicillin production was increased (Fig. 8.6a and b). However, this simple explanation has not been supported by more recent studies in which it has been shown, using continuous culture techniques, that penicillin production is directly related to growth rate (Pirt and Righelato, 1967). These apparently contradictory results can be explained on the basis of catabolite repression. The rapid growth rates in continuous culture were obtained using limiting concentrations of carbohydrate, so catabolite repression would not operate, whereas in the batch fermentation, glucose levels in excess of 3% may be present which will induce

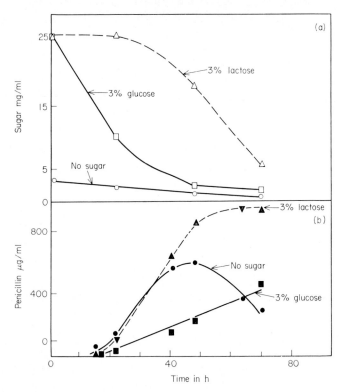

Fig. 8.6. (*a*) Sugar utilization during the penicillin fermentation and (*b*) its effect on penicillin production. From Demain (1968).

catabolite repression. In the batch fermentations it appears to be the conditions normally maintained to stimulate growth rather than growth itself which are incompatible with penicillin formation (Demain, 1968).

Gibberellic acid production. Gibberellic acid, a plant hormone which is used extensively in agriculture and horticulture, is produced commercially from *Gibberella fujikuroi*. In batch fermentations, gibberellin production occurs after the mycelium has reached its maximal dry weight.

Gibberellin synthesis is favoured by a medium which has a low level of nitrogen and a high C/N ratio. It has also been reported that high levels of carbon dioxide, up to 10% of the air volume, stimulate gibberellic acid synthesis (Grove, 1963). Carbon dioxide is known to play a major role in controlling hyphal morphogenesis (Chapter 5) so it is interesting to observe that secondary metabolite production can also be affected by carbon dioxide levels. During the transition from the growth to the production stage, several changes occur in the mycelium; the level of carbohydrate and lipid in the mycelium increases and the rate of protein and RNA synthesis decreases.

Citric acid and gluconic acid production. Although neither citric acid nor gluconic acid are normally considered to be secondary metabolites, their production in industrial fermentations is not associated with growth (Fig. 8.7). This is clearly shown by the use of the pressure

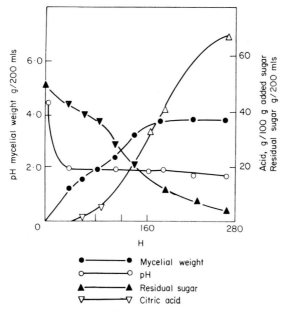

Fig. 8.7. Changes in the pH, residual sugar, citric acid production and mycelial weight during a citric acid fermentation. From Prescott and Dunn (1959).

submerged culture technique for gluconic acid production. In this, the mycelium produced in the initial fermentation is used to transform successive batches of glucose medium into gluconic acid without new mycelium being produced (Fig. 8.8).

In the citric acid fermentation maximal rates of synthesis are not achieved until 160 h at which time the increase in mycelial dry weight has levelled off. It has been demonstrated recently that citric acid production is stimulated by cyclic AMP (see Bu'Lock, 1974). Since in bacteria it has been shown that the level of cyclic AMP is depressed under conditions of catabolite repression and increases when this is reversed, this observation suggests that the catabolite repression may be

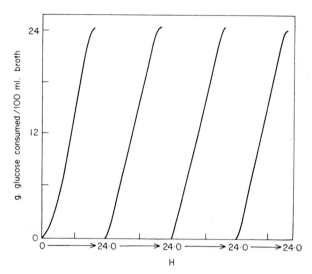

Fig. 8.8. Repeated utilization of *Aspergillus niger* mycelium for gluconic acid production. From Mayer, Umberger and Stubbs (1940).

an important regulatory mechanism in the control of citric acid production.

General Considerations

Since fungi are classified on the form of their reproductive structures, an understanding of the control of reproduction is essential for their identification. Although taxonomy is not the most fashionable of subjects in modern biology, few fundamental or applied studies can progress far without an accurate identification of the organism involved.

In many fungal species, asexual spores appear to be more stable during storage than vegetative mycelium. The production of spores in relatively large quantities is therefore important in the preparation and storage of standardized inocula of highly productive strains obtained in strain selection programmes. Upon discovery of an improved strain, a common procedure consists of growing a sporulating culture on agar or in liquid medium. The spores are then harvested as a suspension and preserved either in a soil tube or as hyophylised samples. Sufficient samples are prepared so that each fermentation or group of fermentations can be initiated from a first or second subculture of the original isolate. This reduces the chance of a strain changing as a result of random mutation during repeated subculturing.

Fungal spores provide a very good form of standardized seed stage. They are reasonably constant in size, readily dispersed and probably possess a more uniform physiological state than a mycelial inoculum. More recently techniques have been developed in which fungal spores rather than mycelium are used to catalyse specific chemical conversions. In these techniques the spores represent packets of enzymes which can be stored, transported, added to reaction vessels or even packed in columns. The phase of biomass production, in this case spore production, can be totally separated from the production phase (Vezina and Singh, 1974).

The capacity to induce both asexual and sexual reproductive stages is desirable for satisfactory strain selection. Unfortunately, many economically important fungi do not have a perfect stage, e.g. *Penicillium chrysogenum, Aspergillus niger*. The technique which has been most widely used for strain selection in such species is that of induced mutation. Using either ultraviolet light or chemical mutagens, attempts have been made to alter the genome of the organism by random gene mutation in such a way as to improve product yield. In such a technique, it is essential that any new strain arising can be isolated as a pure culture, free of contamination from non-mutated cells. This is most easily achieved if the mutagenic treatment is applied to separate uninucleate cells which can then be plated out at such a dilution that each viable cell gives rise to a single discrete colony. In most fungi the only stage in the cell cycle where such cells occur is asexual sporulation.

The selection of auxotrophic markers for use in sexual or parasexual breeding also requires the use of uninucleate spores. In those industrially important organisms which have a perfect stage, sexual breeding is the most satisfactory method of strain selection. This requires the development of special media which promote sexual reproduction. The use of acetate medium for yeast sporulation is well known and the difficulties encountered in obtaining sporulation in many strains of yeast is equally well known to those involved in the study of yeast genetics.

Summary

There has been a dramatic increase in the use of fungi for the production of biomass, antibiotics, enzymes, simple organic compounds and beverages during the past half century. This has led to an increase in the degree of sophistication of the fermentation process, and an increasing demand for a basic understanding of the nature of fungal growth and the relationship between growth and product formation.

The kinetics of yeast growth resemble closely those of bacterial growth; however, the growth kinetics of filamentous fungi are more complex. Since growth is restricted to the apical region of the hyphae, not all the biomass of a mycelium is necessarily involved in growth. For exponential growth to occur, the percentage of mycelium involved in growth must remain constant and the number of hyphal tips must increase at the same rate as the mass of the mycelium. This is achieved by increasing the frequency of branching at higher growth rates. Thus the growth form of a filamentous fungus; the degree of branching, the formation of pellets and the age distribution of the cells can be manipulated by controlling the growth rate using different fermentation techniques.

The biosynthesis of many fungal products is not correlated with growth. Such products have been referred to as secondary metabolites. The economic production of secondary metabolites requires that a maximum proportion of the nutrients available are directed into product formation and a minimum into biomass formation. By limiting the supply of one or more of the essential nutrients growth can be restricted and the non-limiting nutrients diverted into secondary metabolite formation. The terms trophophase and idiophase have been used to describe the periods of maximum growth and maximum product formation.

The physiological conditions which stimulate the transition to idiophase and the biochemical changes which occur during that transition indicate a close similarity between the control of idiophase and of differentiation. In several fungi, the production of secondary metabolites is associated with the development of specific reproductive structures. However in others there is no conspicuous morphological difference between those hyphae which produce secondary metabolites and those which do not. Clearly differences exist at the physiological rather than the morphological level.

The production of fungal spores for mutation studies, for the storage of inoculum, for seed stages and for use as a stable form of biomass represents a more obvious application of fungal differentiation to industrial mycology. The cultivation of mushrooms must, however, remain as the most conspicuous example of the exploitation of fungal differentiation for economic gain.

Recommended Literature

General Reviews

Bu'Lock, J. D. (1961). Intermediary metabolism and antibiotic synthesis. *Advances in Applied Microbiology* **3,** 293–342.

Bu'Lock, J. D. (1967). Essays in biosynthesis and microbial development. "Squibb Lectures". John Wiley and Sons, New York.

Bu'Lock, J. D. (1974). Secondary metabolism in fungi and its relationship to growth and development. *In* "The Filamentous Fungi" 1. Industrial Mycology. Ed. Smith, J. E. and Berry, D. R. Edward Arnold, London. (In press.)

Demain, A. L. (1968). Regulatory mechanisms and the industrial production of microbial metabolites. *Lloydia* **31**, 395–418.

Demain, A. L. (1972). Cellular and environmental factors affecting the synthesis of metabolites. *Journal of Applied Chemistry and Biotechnology* **22**, 345–362.

Prescott, S. C. and Dunn, G. C. (1959). "Industrial Microbiology" 3rd Ed. McGraw Hill, New York, Toronto and London.

Smith, J. E. and Berry, D. R. (1974). Ed. "The Filamentous Fungi". 1. Industrial Mycology. Edward Arnold, London.

Turner, W. B. (1971). "Fungal Metabolites". Academic Press, London and New York.

Vezina, C. and Singh, K. (1974). Transformation of organic compounds by fungal spores. *In* "The Filamentous Fungi". 1. Industrial Mycology. Ed. Smith J. E. and Berry, D. R. Edward Arnold, London. (In press.)

Kinetics

Beran, K. (1969). Analysis of growth and multiplication of yeast cells *Saccharomyces cerevisiae* in continuous culture from the aspect of relative age of cells. *In* "Continuous cultivation of Microorganisms". Ed. Malik, I., Beran, K., Feral, Z., Mark, V., Ricica J. and Smrckova, H. Academic Press, New York and London.

Clutterbuck, A. J. (1970). Synchronous nuclear division and septation in *Aspergillus nidulans*. *Journal of General Microbiology* **60**, 133–135.

Katz, D., Goldstein, D. and Rosenberger, R. F. (1972). Model for branch initiation in *Aspergillus nidulans* based on measurements of growth parameters. *Journal of Bacteriology* **109**, 1097–1100.

Mandels, G. R. (1965). Kinetics of fungal growth. *In* "The Fungi" **1**, 599–612. Ed. Ainsworth, G. C. and Sussman, A. S. Academic Press, London and New York.

Niederpruem, D. J. and Jersild, R. A. (1972). Cellular aspects of morphogenesis in the mushroom. *Schizophyllum commune. Critical Reviews in Microbiology* **1**, 545–576.

Righelato, R. C. (1974). Growth kinetics of mycelial fungi. *In* "The Filamentous Fungi". 1. Industrial Mycology. Ed. Smith, J. E. and Berry, D. R. Edward Arnold, London. (In press.)

Mushroom production

Hayes, W. A. and Nair, N. G. The cultivation of *Agaricus bisporus* and other edible fungi. *In* "The Filamentous Fungi". 1. Industrial Mycology. Ed. Smith J. E. and Berry, D. R. Edward Arnold, London. (In press.)

Smith, J. E. (1969). Commercial mushroom production. *Process Biochemistry*. May. 43–46.

Smith, J. E. (1972). Commercial mushroom production. 2. *Process Biochemistry*. May. 24–26.

Singer, R. (1961). "Mushrooms and Truffles". Leonard Hill (Books), London.

Worgan, J. T. (1968). Culture of higher fungi. *Progress in Industrial Microbiology* **8**, 73–139.

Hallucinogenic fungi

Taber, W. A. (1967). Fermentative production of hallucinogenic indole compounds *Lloydia* **30,** 39–66.

Taber, W. A., Brar, S. S. and Giam, C. S. (1968). Patterns of *in vitro* ergot alkaloid productions by *Claviceps paspali* and their association with different growth rates. *Mycologia* **60,** 806–826.

Vining, L. C. and Taber, W. A. (1963). Alkaloids. *In* "Biochemistry of Industrial Microorganisms", 341–378. Ed. Rainbow, C. and Rose, A. H. Academic Press, London and New York.

Wasson, R. G. (1971). "Soma, Divine Mushroom of Immortality". Harcourt, Brace, New York.

Antibiotic production

Hockenhull, D. J. D. (1963). Antibiotics. *In* "Biochemistry of Industrial Microorganisms", 227–299. Ed. Rainbow, C. and Rose, A. H. Academic Press, London and New York.

Nash, C. H. and Huber, F. M. (1971). Antibiotic synthesis and microbiological differentiation of *Cephalosporium acremonium*. *Applied Microbiology* **22,** 6–10.

Pirt, S. J. and Righelato, R. C. (1967). Effect of growth rate on the synthesis of penicillin by *Penicillium chrysogenum*. *Applied Microbiology* **15,** 1284–1290.

Gibberellin production

Borrow, A., Jeffery, E. G., Kessel, R. N. J., Lloyd, E. C., Lloyd, F. B. and Nixon, I. S. (1961). The metabolism of *Gibberella fujikuroi* in stored culture. *Canadian Journal of Microbiology* **1,** 227–276.

Grove, J. F. (1963). Gibberellins. *In* "Biochemistry of Industrial Microorganisms", 320–340. Ed. Rainbow, C. and Rose, A. H. Academic Press, London and New York.

SUBJECT INDEX